Artificial Intelligence in Smart Cities

Artificial Intelligence in Smart Cities presents an interdisciplinary and comprehensive analysis of how AI re-engineers urban life, governance, and infrastructure. The book explores key thematic areas, including smart government, economy, environment, living, people, cybersecurity, and innovation, complemented by Central European empirical case studies. Each chapter examines the impact of AI technologies—machine learning, natural language processing, and IoT integration—on augmented decision-making, sustainability, health, education, urban resilience, and social inclusiveness.

The book integrates empirical evidence, conceptual reflection, and practical deployment to illuminate how AI can be harnessed for inclusive urban growth. It guides readers through the complex landscape of ethics, economics, and privacy concerns while offering insights on algorithmic fairness, data protection, public trust, and AI innovation. This thoughtful examination advocates for a human-centered and ethical approach to Smart City development, balancing technological advancement with societal needs.

Addressed at researchers, urban planners, policymakers, technologists, and graduate students, this resource provides both policy-driven expertise and analytical capability. Readers will gain valuable perspectives on the convergence of urban futures and Artificial Intelligence, equipping them with the knowledge to navigate the challenges and opportunities of AI-enhanced urban environments while ensuring technology serves the public good.

The book provides a thorough and systematic examination of the role of AI in Smart Cities, covering its applications, benefits, and challenges in various aspects of urban development, including governance, economy, environment, mobility, living, and people.

- Offers an in-depth analysis of the key principles and technologies of AI, as well as its applications in Smart Cities, providing readers with a detailed understanding of the potential of AI to transform urban environments
- Presents case studies of successful AI implementations in Central European cities, providing practical insights and lessons learned for policymakers, urban planners, and business leaders seeking to harness the power of AI to create more liveable, sustainable, and resilient cities
- Adopts an interdisciplinary approach, drawing on insights from urban planning, computer science, engineering, economics, and social sciences to provide a comprehensive understanding of the role of AI in Smart Cities
- Provides a vision for a future where AI is used to create a better life for all citizens and inspires a new wave of innovation and collaboration in the development of Smart Cities

Artificial Intelligence in Smart Cities

Enhancing Urban Life

Radosław Wolniak, Kinga Stecuła and Wieslaw Grebski

CRC Press
Taylor & Francis Group
Boca Raton London New York

CRC Press is an imprint of the
Taylor & Francis Group, an **informa** business

Designed cover image: Shutterstock

First edition published 2026
by CRC Press
2385 NW Executive Center Drive, Suite 320, Boca Raton FL 33431

and by CRC Press
4 Park Square, Milton Park, Abingdon, Oxon, OX14 4RN

CRC Press is an imprint of Taylor & Francis Group, LLC

ISBN: 978-1-032-96021-0 (hbk)
ISBN: 978-1-032-96019-7 (pbk)
ISBN: 978-1-003-58761-3 (ebk)

DOI: 10.1201/9781003587613

Typeset in Minion
by Newgen Publishing UK

Contents

Authors

Radosław Wolniak is a full professor at the Department of Economics and Informatics at the Silesian University of Technology (Poland). For over 20 years he has been dealing with the issues of management, including Industry 4.0, Smart Cities, digitalization and Artificial Intelligence. His achievements include over 500 publications issued by national and international publishers, as well as many undertakings and projects implemented in cooperation with the business environment. Scopus indicators: citations—1321, Hirsh index = 21. He's on the Stanford list of Top 2% scientists most cited across the globe for 2023.

Kinga Stecuła earned a PhD in mechanical engineering at Silesian University of Technology (SUT), Faculty of Organization and Management, Poland, in 2021. Currently, she works in Department of Production Engineering at SUT in the Faculty of Organization and Management. Dr Stecuła deals with research in the field of technical sciences connected with production engineering, but she is also interested in examining human-machine/human-computer interaction and communication. Currently, she is interested in the human-machine relation in the context of virtual reality, augmented reality, and different means of technology. Her interests also include Smart Cities, Industry 4.0, and management. Scopus indicators: citations—250, Hirsh index = 12.

Wieslaw Grebski is Professor Emeritus at the Pennsylvania State University, USA. He was a member of the Engineering Faculty and Coordinator of the Engineering Program at Penn State Hazleton. His research interests evolved from computer-aided manufacturing to sustainable forms of energy. Professor Grebski has always promoted a close cooperation between universities and industry. Under his leadership a new engineering program focusing on renewable energy was developed. He has coauthored 200 publications and 9 books. Professor Grebski promoted unconventional methods for teaching and learning. He received the Milton S. Eisenhower Award, one of the highest recognitions at Penn State. Professor Grebski received 11 other awards for teaching and advising. He has also taught a dual-enrollment class, Introduction to Engineering Design, to promote the engineering profession and recruit students for the engineering program. He was a 15 Information Classification and a general member of the Pennsylvania Academy of Science, the American Society of University Professors, the American Society of Mechanical Engineers, and the American Association of Engineering Educators.

Introduction

The introduction of Artificial Intelligence into city systems is reconfiguring the intellectual and practical foundations of 21st-century cities. Cities are increasingly grappling with every issue from social injustice and climate change to infrastructure decay and mounting pressure on public services, and AI is not an add-on, a marginal overlay, but an integral component of smart urbanism. By no means restricted to technological systems themselves, AI touches on managing cities, the distribution of resources, and the cadence of daily life. This book offers a straightforward, interdisciplinary entry point to the place of Artificial Intelligence in constructing Smart Cities, both its transformative potential and the ethical, social, and political implications that it brings.

At the center of the Smart City concept is a socio-technical imaginary: cities are not merely agglomerations of population and infrastructure but evolving systems governed by flows of data, algorithms, and automated decision-making. The chapters in this volume approach this paradigm from a variety of angles—governance, economy, environment, mobility, living, inclusivity, cybersecurity, innovation, and ethics—so providing an exhaustive description of how AI is implemented and contested in urban environments. Significantly, the book calls for AI in Smart Cities not to be limited by efficiency indicators and infrastructure optimization. Instead, emphasis should be on ensuring that AI technologies are underpinned by sustainability norms, inclusivity norms, transparency norms, and democratic accountability norms.

The book covers a critical analysis of smart governance. In this context, AI is considered a means of institutional agility, citizen participation in decision-making, and responsiveness of policies. Computer-based platforms allow citizens to collaborate with each other to co-design policy, participate in public consultations, and track administration

DOI: 10.1201/9781003587613-1

performance. They can augment levels of transparency and combat bureaucratic lethargy but need new kinds of algorithmic accountability alongside digital literacy among citizens. The use of AI in public administration is defined by a transition from top-down governance structures to networked, data-driven patterns of coordination, with real-time analytics employed to adaptive management of urban complexity.

In an economic sense, AI is already transforming the urban labor market and value creation processes. While AI enhances productivity and drives innovation, especially in industries like logistics, healthcare, and finance, it also accelerates job polarization and technological unemployment. The smart economy therefore has to be considered as a field of tension: AI generates new business models and service delivery opportunities and at the same time generates the potential for displacement for some work. The smart economy chapter highlights reskilling programs, education reform, and public-private-university collaborations to enable technological advances into inclusive economic benefit.

The book also examines the function of AI in mobility systems, the circulatory infrastructure of contemporary urban life. AI enables smart traffic management, autonomous cars, predictive maintenance of transport networks, and integrated multi-modal mobility platforms. These technologies have the potential to reduce congestion, lower emissions, and enhance safety on roads. Their deployment, though, also generates concerns with respect to the governance, access, and digital divide for vulnerable groups like the elderly and the disabled. Of particular concern is how mobility systems supported by AI can be designed to ensure continued inclusivity and avoid widening spatial or social inequalities in the city.

Environmental sustainability is another important aspect of the volume with explicit examination of the ways in which AI aids ecological resilience. Applications include real-time air and water quality monitoring, sound pollution surveillance, and optimization of energy and waste management systems. Smart grids, AI-based forecasting, and adaptive control enable cities to move towards circular and sustainable regimes of resources. Demand can be predicted by systems, environmental externalities can be prevented, and mitigation measures can be enabled in the face of climate change. But the ecological benefits of AI are not automatic—dependent on good system design, fair access to data, and sustained political will.

This volume does more than document the technical potential of AI in the city; it also critically analyzes the normative frameworks, institutional barriers, and ethical conflicts that condition its adoption. Instead of naively embracing Smart Cities as an inevitable development, the book invites readers to think about what we want our cities to be smart about—and on whose behalf. AI is thus reimagined not as an autonomous mission, but as a civic technology whose legitimacy rests on its potential to further the common good.

Radosław Wolniak
Kinga Stecuła
Wieslaw Grebski

CHAPTER 1

The Role of Artificial Intelligence in Smart Cities

1.1 HISTORICAL EVOLUTION OF AI IN URBAN DEVELOPMENT

With the challenges of urbanization posed continuously to cities across the world, Smart Cities have emerged as a very promising solution. In this evolving root lies Artificial Intelligence (AI), in short AI, which is bound to change many facets of urban living. AI makes urban transportation systems more efficient, uses environmental resources more effectively, enhances governance systems, improves the living standards of the residents, develops the economy further, and empowers citizens to take an active interest in the life of the community (McKinsey 2018). Smart Cities put to use new technologies, advanced data analytics, and state-of-the-art digital infrastructure to achieve more efficient, sustainable, and more livable conditions for its citizens (Wolniak and Jonek-Kowalska, 2022; Wolniak et al., 2024; Jonek-Kowalska and Wolniak, 2023).

Advanced urban environments implement a myriad of intelligent solutions across sectors: transport, energy, healthcare, governance, and education. Devices such as IoT, which is required to collect massive amounts of data, are underpinned by integrated digital platforms that allow easy communication and coordination among key stakeholders. The vision of Smart Cities, therefore, runs very closely in the lines of Industry 4.0 and the recent buzz about Society 5.0—a conceptual framework that finds technology playing a driving force toward societal and economic advancements in its core (Maddikunta et al., 2022; Jonek-Kowalska and Wolniak, 2021, 2022).

People have been fighting to be at war for progress, creating cities with more development, environmental awareness, and informed decisions. This is what AI has contributed so much to. AI not only drives technological innovation in smart homes but also influences

DOI: 10.1201/9781003587613-2

human behaviors and lifestyles. Applications are spread in almost all aspects of daily life, giving evidence of its great power to shape the face of future cities (Mourtzis et al., 2022).

AI is a part of this fast-growing science known as computer science that focuses on the development of systems able to perform tasks requiring human intelligence. The list includes reasoning, learning, problem-solving, perception, and understanding language. Its backbone relates to developing machines able to simulate human cognitive processes so as to operate on their own, adjust to new information, and reach conclusions on their own without interference from human beings (Skubis et al., 2024). The ultimate achievement of AI does not lie merely in the duplication of human capabilities but, rather, in their augmentation in ways perhaps unimaginable that will truly revolutionize industries, science, and society (Hole and Ahmad, 2021; Korteling et al., 2021).

AI designs algorithms and computational models that enable machines to perform human-like intelligent tasks. The definition encompasses a wide range of activities, such as simple ones from image identification or speech processing to highly complex autonomous decision-making or strategic planning. AI refers to a comprehensive set of subfields, which make use of machine learning, natural language processing, robotics, and cognitive computing. It does this through the development of systems that can perceive their environment, understand the information, and process it; reason about the information; and finally act in an optimal performance based on goals they have been set for (Pope and Ma, 2024).

AI can be segmented into three major types, each representing a different level and capability (Table 1.1) (Leong, 2021; Banafa, 2023; Ahiwar, 2023):

- Artificial Narrow Intelligence: Also called Weak AI, ANI is a class of AI systems that perform particular tasks. These are really good in performing narrowly defined functions, such as voice recognition, recommendation systems, or playing chess. They cannot generalize beyond their programmed capabilities and cannot do anything other than what they were specifically designed for. Examples of ANI include Siri, Alexa, and Google's search algorithms.
- Artificial General Intelligence: AGI, sometimes referred to as Strong AI, is a more advanced form of AI that really attempts to emulate human cognitive capabilities across a wide variety of tasks. An AGI system would not only excel in specific tasks but also understand context, make decisions in unfamiliar situations, and learn new concepts without human intervention. While AGI is little more than a theory at this juncture, it nonetheless happens to be the general aim of AI studies because that would connote a quantum jump in capability on the part of the machines to mimic the entire gamut of human intelligence.
- Artificial Superintelligence: ASI is the future, hypothetical stage of AI at which machines would be beyond human intelligence in all aspects, including creativity, problem-solving skills, and even emotional intelligence. In this order of AI, it would be able to innovate, adapt, and evolve beyond what humans could do and definitely better at tasks compared to human capability. It has a huge potential and, at the same time huge ethical and existential risks. Actually, it will replace the existing relationship of humans with machines.

TABLE 1.1 The Types of AI

Aspect	Narrow AI (Weak AI)	General AI (Strong AI)	Superintelligent AI
Definition	AI designed for a specific task or narrow set of tasks	AI with the ability to perform any intellectual task that a human can	AI that surpasses human intelligence in every domain
Scope of functionality	Limited to predefined tasks (e.g., facial recognition, voice assistants, recommendation systems)	Can perform a wide variety of tasks and adapt to new, unforeseen problems	Can outperform humans in all intellectual tasks and exhibit creativity
Complexity	Simple, task-specific algorithms and models	Complex systems capable of reasoning, learning, and understanding like humans	Extremely complex, with the ability to improve its own intelligence and capabilities autonomously
Adaptability	Cannot adapt beyond the tasks it is designed for	Can learn, adapt, and solve new problems not explicitly programmed	Continuously learns and evolves, exceeding human-level adaptability
Current status	Fully realized and widely used in various industries	Still theoretical, with no practical implementations	Hypothetical and speculative, with no real-world examples yet
Examples	Siri, Google Search, recommendation systems, self-driving cars	There are no real examples yet, but it would be like a human-level AI	A system that could independently create groundbreaking technologies, solve global problems, and perform tasks beyond human understanding
Key technologies	Machine learning, deep learning, natural language processing, expert systems	Advanced machine learning, cognitive computing, deep reasoning capabilities	Advanced AI systems, recursive self-improvement, potential quantum computing
Ethical and social implications	Lower ethical risks, primarily related to bias, privacy, and misuse	Ethical dilemmas about autonomy, rights, and accountability	Speculative concerns include existential risks, loss of control, and the potential for misuse at an unprecedented scale
Impact on employment	May replace or augment certain job roles but often works alongside humans	Could potentially replace or significantly transform a wide range of professions	Could lead to massive disruption in the job market, with many human roles rendered obsolete
Control and safety	Easily controlled and monitored by humans	Potentially controllable but could raise concerns about autonomy	Hard to control due to its superior intelligence, leading to potential existential risks

Source: Author's own work.

The historical evolution of AI in the context of urban development reflects the exciting tension between innovation in technology and the rapidly changing state of modern cities (Table 1.2). Indeed, this process began in the middle of the 20th century, when computational technologies were first established and AI began to be developed. The role of AI in urban contexts was, at first-stage one, indirect, although entirely academic, belonging in the realm of investigation and theorizing. Most early systems of AI handle mathematics (in the form of problems, optimization processes, and simulations of human cognition)—strong foundations that would later be laid for urban planning and management (Hongladarom et al., 2024).

TABLE 1.2 Evolution of AI in Urban Development

Time Period	Technological Development	AI Application in Urban Development	Challenges and Limitations
Mid-20th century	Early computational systems and theoretical AI research	Limited direct application; the work focused on academic exploration and the solution of simple optimization problems	Limited computational power and lack of urban-specific data
1960s–1970s	Introduction of computer-aided design (CAD) systems	Automated design in urban planning and architecture at high levels of precision	Basic automation without true AI capabilities
1980s–1990s	Emergence of expert systems and GIS	Rule-based AI to aid in decision-making over zoning, transport, and resource allocation, also integrated with GIS to manage spatial data	Rule-based systems lacked adaptability; limited by static data and computational constraints
2000s	Growth of big data and IoT devices	Smart Cities emerged; AI used for real-time traffic signals, optimization of public transportation, and environmental monitoring	Fragmented data sources; challenges in real-time integration
2010s	Advancements in machine learning (ML) and deep learning	Predictive analytics using simulation tools for energy consumption, disaster response, and development assessment for urban sustainability	Risk of biased models and dependency on data quality
2020s–Present	Sophistication of AI in automation, NLP, and personalization	AI-powered climate resilience, affordable housing, and urban equity solutions; autonomous systems like self-driving cars integrated into cities	Ethical concerns, data privacy risks, and the challenge of ensuring inclusivity in AI applications

Source: Author's own work.

Computer-aided design systems, integrated into the 1960s and 1970s, formed a major milestone for urban development. The systems were not "intelligent" per se but signaled a starting point for automation in urban planning and architectural design. Much emphasis was laid on precision and efficiency to provide good groundwork for incorporating complex AI algorithms in the future. These applications of AI in the urban context are actually limited by hardware and data availability (Singh et al., 2024).

That, again, was followed by the dawn of expert systems—those using rule-based logic to model human expertise—in the 1980s and 1990s. These began playing their role in urban planning through a set of decision-support tools that helped with zoning, transportation, and resource allocation. Enhancements with geographic information systems in the same period allowed for the layering of spatial data to visualize and analyze complex relationships in the city. More data-driven approaches to urban challenges began to evolve as early AI and GIS combined.

Therefore, once the 21st century kicked in, big data fueled that exponentially grown computing power. The landscape of AI applications on urban development had absolutely been remade. Cities became places where data is generated: sensors, cameras, or generally what is called IoT devices—everything to render a city capable of evolving into what today may be called Smart Cities. It came as a follow-through of AI-powered systems starting to process all of this in real time for increased urban efficiency. These were the usages of an AI algorithm in traffic congestion solutions, dynamic change in traffic light systems per flow and time, route and schedule optimization using predictive analytics in public transport systems (Dakhole and Praveena, 2024).

Such advancements began in the 2010s, and machine learning is most accurate—deep learning. It breaks away from prior rule-based patterns with the following: dredging massive amounts of data to identify patterns and predict almost literally without human intervention. Applications with many metropolitan spectrums—transport, energy, waste disposal, disaster response, and many more—are growing. Of these, some AI models are primarily created to forecast trends in energy consumption to allow cities to configure their power grids more efficiently and reduce their carbon footprints. Meanwhile, cities supplemented AI with advanced urban modeling tools to simulate how various development projects would affect sustainability and resource allocation (Grudin and Brinkman, 2024; Rasheed et al., 2024; Wang et al., 2023). Recently, AI has entered on an unceremoniously advanced incorporation into urban development: AI is applied by cities to address some of the most pressing challenges in cities today, such as climate change; affordable housing; social equity; and so forth. These Natural Language Processing (NLP) tools can improve active citizenship by diving into the analysis of responses or feedback given by different segments of the population for policymaking. Or it could involve self-running systems within the urban context that would change mobility itself, regardless of traditional infrastructure, like driverless cars. Further, AI services hyper-personalize them by making them tailored to the needs of various communities in an urbanized context (Bassett and Roberts, 2023).

While these are great strides, AI in urban development brings forth a host of challenges from the start: from ethical issues to data privacy and possible perpetuation of bias. In this sense, there is increasing stress on the need for decisions made by AI within urban planning to be more transparent, accountable, and inclusive.

AI in Smart Cities has grown leaps and bounds in the last decade, with widespread adoption, while it is now focused on three critical themes: sustainability, efficiency, and inclusivity (Table 1.3). The AI evolution in urban settings reflects a convergence between technological innovation, increasing data availability, and societal demand for smarter and responsive cities. A decade ago, the concept of Smart Cities was all about deploying Internet of Things (IoT) devices and basic data analytics. AI had a very limited role in optimizing traffic flow using static algorithms or managing energy consumption in specific buildings. No sooner had computational powers and algorithmic sophistication improved than AI reached the fore and began to change the way cities were designed, managed, and experienced (Hongladarom et al., 2024).

Anyhow, this became much evident in the last decade of proliferation of both IoT devices and 5G connectivity to make even unimaginable volumes in real time, thereby turning cities into smart ones unleashing AI applications with newer dimensions of

TABLE 1.3 The Development of AI in Smart Cities over the Last 10 Years

Year	Key Developments in AI for Smart Cities	Technological Advancements	Impact on Urban Life
2014	Initial adoption of IoT in Smart Cities	Basic sensor networks for data collection, early stage of data analytics	Early stages of traffic management and energy optimization through data gathering
2015	Introduction of AI-based traffic management systems	Development of real-time traffic data processing algorithms	Improved traffic flow in pilot cities, reduction in congestion and emissions
2016	Emergence of machine learning algorithms for predictive analytics	Introduction of predictive maintenance for public transportation systems	Reduction in transportation costs and downtime, improved service reliability
2017	Rise of AI in energy management systems	Smart grids powered by AI, machine learning-based energy optimization	Enhanced efficiency in energy distribution, integration of renewable energy sources
2018	Launch of AI-powered surveillance and public safety systems	Development of AI-based computer vision for real-time monitoring of public spaces	Improved public safety, faster emergency response, detection of anomalies in real time
2019	Expansion of autonomous vehicle testing and AI-driven mobility solutions	Increased deployment of AI in autonomous vehicles, smart parking, and transportation networks	Progress toward smarter and safer mobility systems, improved traffic management
2020	Widespread adoption of Natural Language Processing (NLP) for citizen engagement	AI-powered chatbots and virtual assistants, multilingual support	Enhanced citizen-government interaction, improved access to city services
2021	AI-based environmental monitoring systems deployed	AI in satellite imagery analysis for urban planning, air quality monitoring, waste management	Better urban planning, more efficient resource allocation, improved environmental health
2022	Integration of AI across multiple urban systems	Edge computing for faster, local data processing, integration with smart infrastructure	Reduced latency, improved real-time decision-making in traffic, utilities, and services
2023–2024	Holistic AI-powered Smart City ecosystems become more common	Advanced AI algorithms, greater use of 5G, expansion of machine learning and computer vision	Comprehensive improvements in urban sustainability, mobility, safety, and service efficiency

Source: Author's own work.

dynamism and adaptability. For example, intelligent traffic management has moved from simple timed signals to one where AI-driven platforms make use of feeds of live traffic to dynamically modulate the signals with a view to congestion reduction in order to reduce emissions. Probably the most significant change in this period is embedding machine learning with predictive analytics into urban infrastructures. AI Predictive Systems in transport started forecasting demand for public transit, optimizing routes to save costs. The next frontier for applying AI, autonomous driving, moves from theoretical prototypes to pilots over entire cities, demonstrating for human mobility its potential: revolutionary (Foote, 2022).

The character of AI applications for energy management changed. One of the emerging enablers of smart grids for renewable integration, the employment of machine learning algorithms analyzes consumption patterns and provides weather forecasting for efficient energy distribution with less wastage. Energy use is optimized through automation of lighting, heating, and cooling by detecting occupancy and usage in a particular area. AI-powered building management systems automatically went on to optimize such energy use.

Public Safety and Surveillance—AI brought in lots of developments in this respect. A decade ago, all surveillance systems relied on human monitoring. Today, AI-powered computer vision systems analyze video feeds in real time for anomalies such as accidents, crimes, or unauthorized access. These developed into responses to emergencies and improved urban security, yet raised very important questions with respect to privacy and ethical deployment (Haupt et al., 2022).

That is when, as NLP advanced, the right citizen engagement mechanisms for cities emerged. AI chatbots and virtual assistants were incorporated into the system for informing residents of city services, receiving and processing complaints, and facilitating all digital interactions. It integrated multilingual capabilities, thus giving access to the diverse population. Such usages of AI really expanded in urban planning and environmental monitoring. Thanks to AI-based satellite image analysis, cities can monitor deforestation, track air and surface water quality, and control sprawl. It can be a source for decision-making on zoning management and allocation of resources, as well as disaster preparedness (Foote, 2022).

Ethics for AI in Smart Cities started to take center stage probably within the last decade when this AI began integrating deeper into citizens' lives; matters of bias, transparency, and data privacy started to be scrutinized (Lier et al., 2024). It naturally brought about the development of models and policies that would ensure responsible use of AI: cities were working in tandem with academics, technologists, and policymakers to handle the challenge. In other words, with the maturing of AI technology, Smart Cities have gradually moved from isolated applications to comprehensive and interrelated systems, holistic urban ecosystems in which AI harmonizes transport, energy, healthcare, education, and governance into an integrated, sustainable approach to city living (Wolniak and Stecuła, 2024).

1.2 KEY PRINCIPLES AND TECHNOLOGIES OF AI

AI probably constitutes one of the fastest-growing, innovative, and hugely resourceful areas in technology research. It basically constitutes the use of computational processes to simulate human cognition. Basic ideas of AI and the technologies supporting it stem from logical mathematics, computer science, cognitive science, and engineering. AI generally deals with developing systems that are capable of learning, reasoning, perceiving, and acting-things a human would do if he were to perform such activities, which all vary and would therefore require intelligence on his part (Díaz-Rodríguez et al., 2023). The backbone behind AI is machine learning: the design of algorithms which enable computers to learn from data, improve in some context over some kind of time frame, and do so without explicit programming for the job. Machine learning contains two broad variations: one is supervised learning, where the model learns from labeled datasets, and the other is unsupervised learning, where the model finds patterns from an unlabeled dataset. That would be important because it will then enable the system, through reinforcement learning, to come out with optimum behavior patterns through trials and errors using feedback provided in reward/penalty forms (Wahde and Virgolin, 2021).

Deep learning is a subset of machine learning wherein the artificial neural networks are inspired by the human brain, with thousands of layers. In other words, it's made up of layers of interconnected nodes or neurons that process inputs and, subsequently, alter the information. It enables the detection of complex patterns and, subsequently, has enabled state-of-the-art breakthroughs in image recognition, natural language processing, and autonomous systems. AI systems are often reinforced with technologies related to natural language processing, making machines understand, interpret, and create human languages. For this reason, it has a number of beneficial uses in applications related to virtual assistants, chatbots, and automated translations (Wolniak et al., 2024; Stecuła et al., 2024). Similarly, computer vision lets machines analyze visual information; this has become essential for various important tasks like face recognition, object detection, and medical imaging among many others (Simeone et al., 2025).

Some of the key challenges of AI concern how systems reason and come up with efficacious conclusions. Probabilistic reasoning, rule-based approaches, and optimization methods were some of the topics that were covered. Most decision-making models should embed uncertainty and incomplete information, often related to aspects concerning logic and probability. Key technologies underpinning AI include high-performance computing infrastructure (George and Landryova, 2019). Whereas recent breakthroughs in hardware, such as GPUs and TPUs, have really accelerated such complex processes normally involved in training high-end AI models, on their part, cloud computing enables the handling of colossal-size datasets with a heavy computational load. This has helped democratize the application of AI among scholars and industry researchers on every continent (Lier et al., 2024).

Introduction of the concept of ethics in design concerning the AI systems follows. In addition, developers should address concerns about bias, transparency, accountability, and privacy in the development phase of the systems to ensure AI works in the best interest of humans with limited harm. Their success depends upon collaboration among technologists, ethicists, and policymakers in formulating guidelines and regulation that guide the deployment of AI (Mindigulova et al., 2023). It integrated all these principles and technologies, with huge benefits accruing—in a way—to personalized medicine, predictive analytics, autonomous cars, and highly advanced robotics. In any case, for AI to keep progressing, sustained research is required to overcome technological hurdles, enhance model interpretability, and have the goals of AI systems better aligned with those of humans and society at large (Koebe, 2025).

1.3 TRANSFORMATIVE POTENTIAL OF AI IN URBAN ENVIRONMENTS

AI is the key to making these urban cities smart, efficient, and sustainable. These rapid demands are surmounting the efficiency of urban response capability in the rapid era of population growth and climatic change with very limited available resources, while AI opens up new possibilities to handle such complicated issues and support unparalleled development in metropolitan regions (Zheng et al., 2024).

Probably the most significant effect of AI is brought forth with the betterment of infrastructure and services in a city. Every day, massive streams of information flow out from sensors, IoT devices, and public systems (Ponnusamy et al., 2024). AI can digest this real-time data and give actionable insights that enhance operational efficiency. For instance, AI-driven systems anticipate the flow of traffic to optimize public transport timetables or to change traffic lights on the fly and keep clogged-up transportation systems free. Such innovations have improved motility with reduced carbon emissions, hence contributing to environmental sustainability (Pratibha et al., 2024).

Energy management has become indispensable in the urban lifestyle of these modern times, and AI is highly contributing to it. Correspondingly, predictive analytics enables the AI system to forecast the future energy consumption trends that utilities can use to balance up supply and demand effectively. In the same way, integrating with renewable sources of energy, reducing dependencies on burning fossils, and strengthening the resiliency of energy infrastructures are easy for AI-enabled smart grids. AI-based automation of different systems within the building enables the change of lighting, heating, and cooling based on occupancy and atmospheric conditions—all this at considerably reduced operation costs (Stecuła et al., 2024; Wang et al., 2023).

AI in this regard can completely transform the entire aspect of structure in urban planning and building. Advanced modeling software using AI has enabled planners to model the effects that a proposed undertaking will have on traffic flow, environmental impact, and population shifts. The added ability to make more informed decisions relative to long-term sustainability makes it all the more useful. In addition, AI can also be used to determine the most suitable place for infrastructure such as schools or hospitals using demographic and geographic information.

AI has greatly contributed to public safety and its emergencies. Its application in the law enforcement agencies for predictive policing, through data analysis, optimizes resource utilization with a view to spacing out criminal activities (Jogarao et al., 2024). It aids in predicting calamities like floods or earthquakes through space-based intelligence and allows for speedy response supported by real-time data analysis. The enhanced video surveillance system enabled with AI technology allows for better monitoring of city spaces, detection of anomalies in space or events, and overall security (Marvin et al., 2022).

So, actually, the transformative power of AI deepens into public health and social equity of cities. For health, AI would keep assessing the population health trends right from wearables to Electronic Health Records (EHRs), help predict the very next outbreak, and hence intervene and allocate resources in due time. AI in Cities is used to find hotspots of underprivileged areas in cities to feed into targeted policy measures by the improvement of entitlements such as housing, education, and others in issues of equality (Yadav et al., 2025).

AI used in city governance opens completely new perspectives for the involvement of citizens: NLP-based tools analyze citizens' feedback, which is gathered from social media, public debates, and questionnaires, hence giving politicians greater depth of people's

TABLE 1.4 The Transformative Impact of AI in Smart Cities

Sector	AI Applications	Transformative Impact	Examples
Smart infrastructure	AI-powered sensors, predictive maintenance, automated building systems, smart grids	Optimizes energy usage and reduces waste, increases lifespan and reduces costs of urban infrastructure, enhances safety with real-time monitoring and quick response	Smart meters in energy grids, AI-based building management systems, predictive maintenance of bridges and roads
Mobility and transport	Autonomous vehicles, traffic management systems, AI-powered public transport, route optimization	Reduces traffic congestion, enhances transportation safety and efficiency, minimizes environmental impact by optimizing routes and vehicle usage	Self-driving cars and buses, AI in smart traffic lights, real-time public transport scheduling and dynamic routing
Urban planning	AI-based urban design tools, data-driven planning, simulation of city development scenarios	Facilitates better space utilization, supports sustainable urban growth, helps in creating livable, accessible, and resilient cities	Urban simulation software for traffic, zoning, and environmental impact, AI for predictive urban growth analysis
Healthcare	AI diagnostics, telemedicine, predictive health analytics, robotic surgery, personalized medicine	Improves healthcare access and efficiency, enhances disease prevention and early detection, personalizes treatment based on individual health data	AI-driven diagnostic tools (e.g., radiology), telehealth platforms powered by AI for remote consultations, robot-assisted surgeries
Sustainability	AI in waste management, water conservation, energy efficiency, renewable energy integration	Reduces waste and pollution, optimizes resource consumption, supports cities in achieving sustainability goals	AI-based waste sorting systems, smart irrigation systems for water conservation, AI in energy grid optimization and integration of renewables
Public safety	AI surveillance, crime prediction, disaster management systems, emergency response optimization	Increases public safety by predicting and preventing crime, speeds up emergency responses, provides better disaster preparedness and management	AI-powered CCTV cameras for real-time threat detection, predictive policing algorithms, AI for emergency response coordination
Economic development	AI for job matching, economic forecasting, market analysis, business innovation	Enhances employment opportunities through improved job matching, supports economic growth by identifying emerging sectors and needs, facilitates entrepreneurial growth	AI-powered platforms for job market insights, AI tools for analyzing economic trends and supporting local businesses, AI in financial services
Social services	AI in social care, personalized education, legal assistance, public welfare programs	Improves access to services for vulnerable populations, personalizes care and education, supports social equity by optimizing resource distribution	AI in personalized learning platforms, AI chatbots for legal advice, social welfare distribution optimization using AI

TABLE 1.4 (Continued)

Sector	AI Applications	Transformative Impact	Examples
Governance and civic engagement	AI in citizen engagement, smart voting systems, predictive policymaking, government transparency	Promotes participatory governance, enhances decision-making with data-driven insights, increases transparency and trust in government systems	AI-driven platforms for public consultations, smart voting systems, AI tools for analyzing public sentiment and policy impacts
Environmental monitoring	AI-based environmental sensors, climate modeling, biodiversity monitoring, air and water quality analysis	Enhances the ability to monitor and protect natural environments, helps mitigate and adapt to climate change, promotes conservation efforts and better environmental stewardship	AI sensors for real-time air quality monitoring, predictive models for climate change impacts, AI to track endangered species and biodiversity

Source: Author's own work.

perceptions. In addition, chatbots and AI-run assistants raise the level of communication between citizens and public authorities, which enhances the effectiveness and transparency of service delivery (Garg et al., 2024).

The transformative potential of AI in cities, despite this, remains high, while challenges would still abound. There are issues related to data privacy, algorithmic bias, and displacement of the workforce that would need to be taken care of if the implementation of AI systems is to happen in an ethical and inclusive manner. Of these, three important risk mitigants are well-designed transparency mechanisms in governance, stringently applied data protection protocols, and stakeholder consultations in continuance (Ali et al., 2025).

Its huge potential would change urban living in terms of infrastructure, public safety, healthcare, and sustainability. Advancement in Breakthrough AI technologies means that cities will create improvements in efficiency, livability, sustainability, and equitability. Integration of AI in urban infrastructure would involve optimization of resources, enhancing the living standards of citizens, and reducing impacts on the environment while innovating in government or public service delivery (Table 1.4) (Aljarrah, 2024; Ojha and Nikhil, 2024).

BIBLIOGRAPHY

Ahirwar, A. (2023). Introduction of AI in innovative engineering. In: A. Ahirwar, P. K. Shukla, M.K. Shrivastava, P. Maheshwary, & B. Gour (Eds.), *Innovative Engineering with AI Applications*. Massachusetts, USA: Wiley, pp. 1–22.

Ali, J., Kumar Singh, S., Jiang, W., Ibrahim Daradkeh, Y., and Mehmood, A. (2025). A deep dive into cybersecurity solutions for AI-driven IoT-enabled smart cities in advanced communication networks. *Computer Communications*, 229, 108000.

Aljarrah, E. (2024). AI-based model for prediction of power consumption in smart grid-smart way towards smart city using blockchain technology. *Intelligent Systems with Applications*, 24, 200440.

Banafa, A. (2023). *Transformative AI: Responsible, Transparent, and Trustworthy AI Systems*. New York: River Publishers, pp. 1–156.

Bassett, C., and Roberts, B. (2023). Automation anxiety: A critical history – the apparently odd recurrence of debates about computation, AI and labour. In: S. Lindgren (Ed.), *Handbook of Critical Studies of Artificial Intelligence*. Boca Raton, Florida, USA: CRC Press (Taylor & Francis Group), pp. 79–93.

Dakhole, D., and Praveena, K.N. (2024). History and role of AI in healthcare and medicine. In: B. Chander, K. Guravaiah, A.B., G. Kumaravelan (Eds.), *Handbook of AI-Based Models in Healthcare and Medicine: Approaches, Theories, and Applications*. Boca Raton, Florida, USA: CRC Press, pp. 19–31.

Díaz-Rodríguez, N., Del Ser, J., Coeckelbergh, M., Herrera-Viedma, E., and Herrera, F. (2023). Connecting the dots in trustworthy Artificial Intelligence: From AI principles, ethics, and key requirements to responsible AI systems and regulation. *Information Fusion*, 99, 101896.

Foote, K.D. (2022). The history of machine learning and its convergent trajectory towards AI. In: S. Carta (Ed.), *Machine Learning and the City: Applications in Architecture and Urban Design*. Hoboken, New Jersey, USA: Wiley, pp. 129–142.

Garg, K.D., Kaur, P., and Sharma, P. (2024). AI-enabled smart homes and buildings in smart cities. In: B. Sharma, M. Gupta, & G. Jeon (Eds.), *Smart Cities: Blockchain, AI, and Advanced Computing*. Boca Raton, Florida, USA: CRC Press, pp. 188–212.

George, G.A., and Landryova, L. (2019). Interaction between human and AI systems – When automated systems move towards autonomous. *Proceedings of the 2019 20th International Carpathian Control Conference*, ICCC 2019, 8765952.

Grudin, J., and Brinkman, D. (2024). HCI history and the trajectory to generative AI. *Conference on Human Factors in Computing Systems – Proceedings*, 597.

Haupt, S.E., Gagne, D.J., Hsieh, W.W., Tissot, P., and Williams, J.K. (2022). The history and practice of AI in the environmental sciences. *Bulletin of the American Meteorological Society*, 103(5), E1351–E1370.

Hole, K.J., and Ahmad, S. (2021). A thousand brains: Toward biologically constrained AI. *SN Applied Sciences*, 3(8), 743.

Hongladarom, S., van der Vaeren, A., and Sakprasert, S. (2024). The history of AI in Thailand: Thickening our vision of AI by caring for a marginalised actor. *IFIP Advances in Information and Communication Technology*, 719, 23–45.

Jogarao, M., Lakshmanna, B.C., and Naidu, S.T. (2024). AI-enabled circular economy management for sustainable smart cities: Integrating artificial intelligence in resource optimization and waste reduction. In: V. Kandpal, E.D.R. Santibañez-González, P. Chatterjee, & M.K. Nallapaneni (Eds.), *Smart Cities and Circular Economy: The Future of Sustainable Urban Development*. Bingley, United Kingdom: Emerald Publishing, pp. 83–96.

Jonek-Kowalska I., and Wolniak R. (2021). Economic opportunities for creating smart cities in Poland: Does wealth matter?. *Cities*, 114, 103222, 1–16.

Jonek-Kowalska I., and Wolniak R. (2022). Sharing economies' initiatives in municipal authorities' perspective: Research evidence from Poland in the context of smart cities' development. *Sustainability*, 14(4), 2022, 2064, 1–23.

Jonek-Kowalska, I., and Wolniak, R. (2023). *Towards Sustainability and a Better Quality of Life?* Routledge: London, UK.

Koebe, P. (2025). How digital technologies and AI contribute to achieving the health-related SDGs. *International Journal of Information Management Data Insights*, 5(1), 100298.

Korteling, J.E., van de Boer-Visschedijk, G.C., Blankendaal, R.A.M., Boonekamp, R.C., and Eikelboom, A.R. (2021). Human- versus Artificial Intelligence. *Frontiers in Artificial Intelligence*, 4, 622364.

Leong, B. (2021). *General and Narrow AI, Encyclopedia of Artificial Intelligence: The Past, Present, and Future of AI.* California, USA: ABC-CLIO, pp. 160–162.

Lier, S.K., Gerlach, J., and Breitner, M.H. (2024). What is ethical AI? – Design guidelines and principles in the light of different regions, countries, and cultures. *Proceedings of the Annual Hawaii International Conference on System Sciences*, pp. 6848–6857.

Maddikunta, P.K.R., Pham, Q.-V., Nguyen, D.C., Huynh-The, T., Aouedi, O., Yenduri, G., Bhattacharya, S., and Gadekallu, T.R. (2022). Incentive techniques for the Internet of Things: A survey. *Journal of Network and Computer Applications*, 206, 103464. https://doi.org/10.1016/j.jnca.2022.103464

Marvin, S., While, A., Chen, B., and Kovacic, M. (2022). Urban AI in China: Social control or hyper-capitalist development in the post-smart city?. *Frontiers in Sustainable Cities*, 4, 1030318.

McKinsey Global Institute. (2018). *Smart Cities: Digital Solutions for a More Livable Future.* McKinsey & Company.

Mindigulova, A.A., Vikhman, V.V., and Romm, M.V. (2023). The use of artificial intelligence in education: Opportunities, limitations, risks. *International Conference of Young Specialists on Micro/Nanotechnologies and Electron Devices*, EDM, 2023-June, pp. 2000–2003.

Mourtzis, D., Angelopoulos, J., and Panopoulos, N. (2022). A literature review of the challenges and opportunities of the transition from Industry 4.0 to Society 5.0. *Energies*, 15(18), 6276. https://doi.org/10.3390/en15176276

Ojha, N., and Nikhil, V.P. (2024). Challenges of AI implementation for boosting circular economy in smart cities. In: V. Kandpal, E.D.R. Santibanez-Gonzalez, P. Chatterjee, & M.K. Nallapaneni (Eds.), *Smart Cities and Circular Economy: The Future of Sustainable Urban Development.* Bingley, United Kingdom: Emerald Publishing, pp. 215–233.

Ponnusamy, S., Chourasia, H., Rathod, S.B., and Patil, D. (2024). Advanced computing for smart public transportation systems in smart cities. In: B. Sharma, M. Gupta, & G. Jeon (Eds.), *Smart Cities: Blockchain, AI, and Advanced Computing.* Boca Raton, FL: CRC Press. https://doi.org/10.1201/9781003442660-10

Pope, A., and Ma, R. (2024). Exploring historians' critical use of generative AI technologies for history education. *Proceedings of the Association for Information Science and Technology*, 61(1), pp. 1071–1073.

Pratibha Kaur, G., Dev, P., and Kaur, A. (2024). Smart cities unveiled: The AI revolution in urban development. In: A.K. Rana, S. Chakraborty, P. Goel, S. K. Rana, & A.A. Elngar (Eds.), *Internet of Things and Big Data Analytics-Based Manufacturing.* Boca Raton, FL: CRC Press, pp. 167–180.

Rasheed, K., Zaland, A., Saad, S., Ammad, S., and Rostami, A. (2024). History of AI. In: S. Saad, S. Ammad, & K. Rasheed (Eds.), *AI in Material Science: Revolutionizing Construction in the Age of Industry 4.0.* Boca Raton, FL: CRC Press, pp. 15–46.

Sharma, B., Gupta, M., & Jeon, G. (Eds.). (2024). AI-driven traffic management systems reducing congestion and improving safety in smart cities. In: *Smart Cities: Blockchain, AI, and Advanced Computing.* Boca Raton, Florida, USA: CRC Press/Taylor & Francis, pp. 96–121.

Simeone, D., Marchionni, C., and Rotilio, M. (2025). Artificial Intelligence and lean construction: Where are we and where are we going?. *Lecture Notes in Civil Engineering*, 611 LNCE, pp. 661–677.

Singh, R., Singh, S., and Singh, A. (2024). Landmark events, major trends, and the future of AI research: History, promise, hype. *2nd International Conference on Artificial Intelligence and Machine Learning Applications: Healthcare and Internet of Things*, AIMLA 2024.

Skubis, I., Wolniak, R., and Grebski, W.W. (2024). AI and human-centric approach in smart cities management: Case studies from Silesian and lesser Poland voivodships. *Sustainability*, 16(18), 8279.

Stecuła, K., Wolniak, R., and Aydın, B. (2024). Technology development in online grocery shopping—from shopping services to virtual reality, metaverse, and smart devices: A review. *Foods*, 13(23), 3959.

Stecuła, K., Wolniak, R., and Grebski, W.W. (2023). AI-driven urban energy solutions—from individuals to society: A review. *Energies*, 16, 7988.

Wahde, M., and Virgolin, M. (2021). The five is: Key principles for interpretable and safe conversational AI. *ACM International Conference Proceeding Series*, pp. 50–54.

Wang, C., Cai, J., Gao, C., and Ye, X. (2023). History, status, and development of AI-based learning science. *SN Computer Science*, 4(3), 316.

Wang, P., Song, W., Zhou, J., Tan, Y., and Wang, H. (2023). AI-based environmental color system in achieving sustainable urban development. *Systems*, 11(3), 135.

Wolniak, R., Gajdzik, B., Grebski, M., Danel, R., and Grebski, W.W. (2024). Business models used in smart cities—theoretical approach with examples of smart cities. *Smart Cities* 7, 1626–1669.

Wolniak, R., and Jonek-Kowalska, I. (2022). The creative services sector in Polish cities. *Journal of Open Innovation: Technology, Market, and Complexity*, 8, 17.

Wolniak, R., and Stecuła, K. (2024). Artificial intelligence in smart cities—applications, barriers and future directions: A review. *Smart Cities*, 7, 1346–1389.

Wolniak, R., Stecuła, K., and Aydın, B. (2024). Digital transformation of grocery in-store shopping-scanners, artificial intelligence, augmented reality and beyond: A review. *Foods*, 13(18), 2948.

Yadav, A.K., Sharma, A.K., Patel, M., and Rathore, N.S. (2025). Intelligent urban futures: Exploring AI and machine learning applications in smart cities. *Lecture Notes in Networks and Systems*, 1075 LNNS, pp. 335–345.

Zheng, X., Ma, Z., and Yuang, Z. (2024). Urban design and pollution using AI: Implications for urban development in China. *Heliyon*, 10(18), e37735.

CHAPTER 2

AI in Smart Governance

2.1 AI-ENHANCED DECISION-MAKING PROCESSES

The organizational, governmental, and institutional models of governance in contemporary settings have been through radical metamorphoses with the urge to adapt and include factors such as inclusiveness, transparency, and ready ability to adapt. As such, these increasingly have to address arising issues relating to the increasingly complex and changing environment within which new and rapidly developing digital technologies, processes of globalization, and social expectations keep on shifting. In practice, traditional hierarchical systems are being wholly or partly replaced by far more decentralized processes to enable room for a wide range of stakeholders' contributions (Castillo-Villar, 2021). Several models of governance now in vogue include democratic governance, networked governance, corporate governance, and participatory governance, each serving a purpose in particular contexts. However, as far as the models of governance are concerned, significant challenges lie ahead with regard to their effectiveness and efficiency (Downes et al., 2018).

The important challenges of the contemporary governance models are embedded in the decision-making complexity. Since organizations and governments have to deal with increasingly complex, interconnected problems, decisions often involve many actors with varied, often conflicting interests (Table 2.1). This may suggest a slow pace of decision-making, where compromises dilute policy or strategic effectiveness. Moreover, the continuous evolution of technology and markets requires agility and fluidity in governance models, too, which is very hard to realize, considering that systems and processes are usually built on traditional, rigid frameworks.

Another challenge involves the requirement for transparency and accountability in such complex, multi-stakeholder environments. Tracing decision-making, valuing

DOI: 10.1201/9781003587613-3

TABLE 2.1 The Potential of AI to Address the Challenges of Traditional Governance Models

Challenge	How AI Can Help
Inefficiency in decision-making	While most traditional models of governance rely on sluggish and manual processes for decision-making, responses against urban issues are either not seen or have been rather late. AI works with huge volumes of data in real time to present fast and informed decisions. It automates the analysis of data and gives actionable insights to ensure that decisions can be made promptly, hence enhancing overall efficiency in governance.
Limited access to real-time data	Traditional governance systems depend on data, which is usually outdated or limited, and hence cannot address the current issues correctly. AI systems, especially those integrated with IoT devices, are able to provide continuous real-time data from different sources, like traffic sensors and environmental monitors, to the decision-makers for updated information to make proactive decisions.
Resource allocation inefficiency	Most of the traditional governance models cannot utilize their resources with effectiveness, often leading to wasted or underutilized resources. Artificial Intelligence-powered systems guarantee the optimal management of those through their insights about consumption patterns, with future estimates. Practical applications will span from smart distribution systems to waste collection schedules to efficient water usage—all that the resources go precisely where they will be needed most.
Lack of predictive capabilities	Traditional models of governance are normally reactive—that is, they act when the problem has already occurred. AI brings predictive analytics to governments so that they can predict problems in traffic congestion, energy shortages, or public health. This foresight enables proactive measures that reduce the impact of future challenges.
Difficulty in managing complex data	Traffic, pollution, and the use of public services are just part of the complexity with which urban data can overwhelm traditional governance systems. AI works very well with large datasets containing complexity in variables. With machine learning and data analytics, AI will be able to find patterns and correlations in urban data that might be very hard for a human analyst to discover, hence driving better decision-making.
Limited public engagement	Traditional models of governance fall short in their ability to involve citizens effectively in the decision-making process, hence disengagement or dissatisfaction is articulated. AI-driven platforms like chatbots, virtual assistants, and sentiment analysis can help in better communication between citizens and officials. Similarly, AI may analyze opinions on social media and through surveys to make governance processes more inclusive and responsive.
Slow response to emergencies	Traditional governance relies on manual coordination for emergency responses; this usually leads to delay and inefficiency. AI helps in smoothing the process of emergency management through real-time data analysis, which aids in the prediction of potential risks and optimization of response strategies. Such examples include the use of AI in predicting natural disasters, distributing emergency resources more appropriately, and handling evacuations.
Lack of personalization	The traditional model of governance falls flat in trying to meet the plurality of needs coming from citizens or communities. AI does that, however, by delivering customized services where recommendations and solutions are patterned around each individual particular need. Examples include AI providing an optimization of health services depending on personal health records of a certain person, or public transportation depending on routes and timing determined by citizen demand.
Complexity in urban planning	Traditional urban planning can hardly ever simulate the long-term impact a project will have and often leads to unsustainable developments. AI allows for extended simulations and predictive models based on many factors, such as population growth, environmental impact, needs on infrastructure, and so on. The models yield insights that permit urban planners to make informed decisions in land use, zoning, and infrastructure development.
Security and privacy concerns	Traditional models of governance cannot have mechanisms which may potentially ensure data privacy in the case of sensitive information. Security can be improved by providing encryption through AI, conducting anomaly detection, and carrying out surveillance; at the same time, this can also enable privacy by anonymization or secure management of personal data, responding to data protection concerns.

Source: Author's own work.

outcomes, or finding out motivations would be burdensome for citizens or employees in general. General absence of transparency would spur mistrust and estrangement from the process itself, weakening legitimacy of governance structures. Accountability mechanisms are usually poorly developed or poorly enforced-legal provisions apart from ethical concerns, corruption, and inefficiency.

The second major restraint concerns the power of governance systems to cope with the quantity and speed of data emanating from modern technologies. The challenge now before the governance body is not only how to interpret Big Data, Artificial Intelligence (AI), and IoT, but also to react in a timely manner upon receipt. Most of the governance models have collapsed or have failed to readjust in this challenge in light of underutilizing data on occasions when proactive intervention is possible, or vice versa, overstepping with regard to the collection of data, bringing about concerns over privacy (Bosco et al., 2024).

AI, in every respect, points to ways through which such challenges could be solved. Among many promising applications of AI in governance, the one place it is touted as promising is in improving decision-making. The ability of the AI system, especially machine learning algorithms, to process big datasets faster than humans, identifying patterns or trends that might otherwise have been missed, thus leads to improved decisions (Skubis et al., 2024). This brings more precise and timely information to the policymakers, thus helping the latter to make better decisions with a basis on evidence-based policy (Chiordi et al., 2022; Wu, 2022). AI has the ability to model different policy options and their likely impacts for more effective scenario planning and risk assessment. AI provides more enhancements in the area of transparency and accountability because it will provide tools for real-time monitoring and assessment of the governance process (Barragán Vargas et al., 2022). For example, AI-powered systems can monitor project execution and compliance, underlining any anomalies that may indicate corruption or inefficiency. The next step from here, logically, would be the application of such systems in designing even more open, accessible platforms for citizens' engagement with governance, including a view on the processes and outcomes of decision-making not possible before (Dong and Liu, 2023; Fabregue, 2024).

AI helps governance models turn out to be more adaptive, automating all kinds of routine tasks and procedures. This liberates human resources for dealing with more strategic or complex problems, hence making possible the reaction of the governance structures far quicker in light of newly arising challenges (Caprotti and Liu, 2022). Examples might include using AI in corporate governance for continuous monitoring of financial transactions in search of discrepancies so that fraudulent activities, or some other type of misbehavior, would not occur without well-timed management. In government, AI can be used to analyze data on public health, predict trends, better manage resources, and improve the capability to react in cases of pandemics or natural catastrophes.

AI can promise a strong boost to the decision-making process within the structure of Smart Cities, further pushing efficiency and data-driven adaptive performances of governance (Ashwini et al., 2022). Therefore, with the potentiality to undertake real-time processing of vast volumes of data, insights generated within AI will inform city administrator decisions for upgraded urban planning, resource management, and public services (Davide, 2021).

Various ways in which AI helps in decision making are multidimensional. It will also be responsible for processing and interpreting inflow data from various sensors and different equipment installed within the city (Ullah et al., 2023; Pribadi et al., 2023). Screening of traffic, air quality, energy consumption, usage of public transportation—down the line—all these churn out a volume of Big Data that no human operator can efficiently analyze without the power of advanced algorithms (Oleksy et al., 2023). The trends and anomalies it identifies, together with forecasts of upcoming events in light of the historical record, would make it easier to take certain steps to solve an impending problem before it unfolds. That is to say, using machine learning algorithms, AI would forecast congestion, problems, or bottlenecks in traffic and, accordingly, automatically advise on changes in settings for the traffic lights, for instance, or foresee surges in demand on the power grid and hence optimize it (Bokhari and Myeong, 2022, 2023).

AI will drive further the better decision-making regarding predictive maintenance of basic infrastructure. With sophisticated machine learning algorithms, AI systems analyze sensor data from bridges, roads, and water pipelines for the first signs of deterioration. Here, municipalities are able to schedule repairs and replacements long before any actual failures have occurred, averting rather costly failures and enhancing public safety. Similarly, AI-powered monitoring systems enhance disaster management by predicting weather patterns and spotting the most vulnerable spots in infrastructure, hence adopting quicker and more effective emergency response strategies (Jain et al., 2021).

It further makes AI useful in enhancing resource allocation in cities by analyzing consumption and waste patterns. For example, AI systems are improving the water distribution network through leak detection, hence proffering optimal routes that maintenance crews must follow. These algorithms analyze waste collection scheduling and filling levels of the collection containers; subsequently, the route design for pickups could be developed in such a way as to even achieve some savings on fuel and manpower, simultaneously decreasing the negative impact of those activities on the environment (Arora et al., 2023).

AI has been changing public safety and law enforcement on a wide scale by analyzing data from surveillance for suspicious activities. Computer vision amalgamated with machine learning technologies thus enabled the facial recognition system to trace the missing persons or track the individual involved in criminal activities. Further, it provides protection on privacy through anonymization of data when required and balances security with rights of an individual (Al-Besher and Kumar, 2022).

It might be useful in the field of urban planning and development through simulated projections to show what certain proposed initiatives may look like, showing implications concerning traffic flow, level of pollution, and population density. For this reason, simulations will help planners weigh options and choose viable strategies. AI-driven geographic information systems make visualization possible for decision-makers so they get the right orientation in land use, transport network, and green spaces that come forward to achieve both ecological and social objectives (Zhou, 2022).

Other key applications of AI in decision-making have to do with citizen participation through participatory governance. Community residents are in contact with city officials more than ever before, thanks to AI-powered chatbots and virtual assistants (Stecuła et al., 2024). This reduces the response time taken toward queries and complaints. Analytics tools study feedback from public opinion on social media to understand the trend, and thus authorities adjust their policies (Guarda et al., 2023). AI-driven platforms are also creating avenues for online voting and surveys whereby citizens can constructively contribute to decisions taken in respect of urban development initiatives. Added to that, another advantage working in favor of AI-driven decision-making is the processing and interpretation of data, which can even be in real-time. Modern Smart Cities generate a huge amount of information emanating from sources such as IoT sensors, surveillance cameras, traffic management systems, and social media platforms. It is this information that the AI algorithms analyze for patterns, predict trends, and find potential problems before they become critical. Predictive analytics enable the smoothing of traffic flow by analyzing the movement patterns of vehicles and, in turn, adjusting the traffic light signals for reduced congestion and emission. Similarly, AI incident-report analysis contributes to improved emergency response mechanisms through better resource dispatching (Singh et al., 2023). Besides, AI contributes to operational efficiency, hence it is sustainable and resilient. Analyzed energy consumption data by means of AI may allow elaboration on strategies for energy conservation, integrating renewable sources of energy (Li et al., 2023; Stecuła et al., 2023).

AI models will monitor air quality and water use, thereby enabling authorities to take remedial measures with a view to protecting the health of citizens and the natural environment (Jonek-Kowalska and Wolniak, 2023). Equally, these technologies have become very important in disaster management through early warnings and responses in the case of either natural or artificial disasters (Sienkiewicz-Małyjurek, 2023; Saadah, 2021). AI systems support the development of better processes relating to citizen engagement and participation. For example, AI-driven Smart City platforms analyze the general response of the public in terms of needs at the community level and can even prioritize projects based on votes from the residents themselves. Further, this enhances the ease of access through chatbots and virtual assistants that instantly answer informational queries by guiding users through different municipal services. This makes for greater accountability and more transparency in the system of governance (Kuguoglu et al., 2021).

2.2 TRANSPARENCY AND CITIZEN ENGAGEMENT THROUGH AI

Transparency and citizen involvement in today's governance, aided lately by integrations involving AI, have only served to change faces on facets unimaginable. Some new tools and platforms with which these AI technologies feature may make active participation from citizens even deeper inside decision-making processes, while expanding also the accountability and openness of governing structures. Examples include: participatory budgeting platforms (Schintler and McNeely, 2022). These systems use AI to process public input in a manner that guarantees effective resource planning by enabling the citizens to propose and vote on local projects. It democratizes budgetary decisions, streamlining the funds in accordance with public priorities and thereby instilling trust between governments and peoples (Table 2.2).

AI-driven feedback portals develop this aspect of transparency to the fullest by processing and analyzing opinions and concerns mentioned by the citizens. These platforms use natural language processing and sentiment analysis to show trends and issues that governments can act on in due time (Vrabie, 2022). This somehow makes governance more responsive. On the other hand, open data platforms driven by AI have increasingly

TABLE 2.2 Citizen Engagement through AI

Example of Citizen Engagement	Description
Participatory budgeting platforms	AI systems analyze the input from the public to assign funding for community projects transparently. The citizens can provide proposals and vote on priorities for funding.
AI-driven feedback portals	Artificial Intelligence enables the platforms to collect and analyze comments of the citizens, thereby enabling the authority officials to keep track of the trend in the attitudes of the public and effective action.
Open data platforms	Government Data Organization and Visualization: AI helps in the organization and presentation of government data; because of this, government information on budgets, policy, and public services would be even more accessible and understandable by citizens.
Crowdsourcing platforms for policy input	AI tools synthesize and collate public suggestions and feedback on policy decisions to make governance inclusive.
Chatbots for public services	AI-powered chatbots respond instantly to every query of citizens related to any government program or services, hence increasing access and engagement.
AI-assisted surveys and polls	Active AI systems study the response of surveys and public polls to estimate the opinion of the people on certain issues and trends that may affect policy decisions.
Civic hackathons	Events such as hackathons on AI platforms, where a group of people collaborate in finding innovative solutions for problems affecting the community, ensure community participation in such an advanced way.
Smart complaint management systems	AI-powered technologies facilitate sorting and prioritizing citizens' complaints for quicker resolution, hence enforcing accountability.
AI-supported town hall meetings	AI-powered virtual meetings now come with real-time translation, summarization, and sentiment analysis, thus engaging a wider audience through better clarity in communication.
Predictive analytics for urban planning	AI goes ahead to make growth predictions and infrastructural needs a city might encounter using public contributions and data for city planning that truly reflects the will of the citizens.
Crowdsourced monitoring systems	It empowers active citizen contribution of data via mobile applications, which-when analyzed by AI systems-feed into real-time issue monitoring, such as pollution, traffic flow, and safety, back to policymakers.
AI-based voting assistance tools	AI is helping citizens to vote by explaining policies and candidate positions, increasing voter education and participation.
Citizen science platforms	AI will allow large-scale collection and analysis of data, hence allowing the participation of citizens in scientific research and environmental monitoring.
AI ethics and transparency panels	AI has made it possible to create panels that can debate and assess ethics in the use of applications, building accountability and trust therein.
Social media monitoring for civic issues	AI scans social media trends and conversations to identify emerging public concerns that require timely responses by the government.

Source: Author's own work.

made the governments more transparent. These arrange and visualize big chunks of data from the government and show them in understandable format before the citizens. From budgetary allocations to the performance of policies, these platforms have provided avenues for the general public to question government activities and hold them accountable (Harnal et al., 2022).

Citizen participation is also deepened through crowdsourcing of policy input. AI tools collate public suggestions, identify consistent themes, and distill feedback into recommendations that inform policymaking. This is representative and captures a fair share of diverse perspectives, hence equity in governance. AI-powered chatbots and virtual assistants further ease access to public services (Xiao et al., 2021). They immediately inform citizens about various government programs and policies and eliminate all barriers to participation in them. They are most useful in reaching the strata of society that were otherwise unreachable and keep them apprised about basic information. AI-driven surveys and polls are another avenue through which public opinion can be ascertained (Saptadi et al., 2023). These systems scan survey responses to identify trends and major issues to be taken up for policy formulation.

Predictive analytics may enable governments to look ahead in time, foresee challenges, and prepare themselves to meet the challenges by matching the needs and aspirations of citizens. Civic hackathons, organized for AI in support, firmly allow citizens to come together in finding solutions to local problems. Events like these drive and create community ownership of the public initiative. In the same vein, smart complaint management systems flag and address grievances using AI, hence improving service delivery and building confidence (Ng et al., 2023). AI-powered virtual town hall meetings are now commonplace. Real-time translation and sentiment analysis further enhance participation. This has brought in meetings beyond geographical distances and languages to achieve much more inclusive dialogues, making communication transparent. AI in urban planning enhances it further with predictive analytics from the input given by the citizens and from data trends. In this way, the developmental projects keep consideration with the need and priority of the general public, and growth will be for the real improvement of living standards (Pahuja, 2022).

Other applications are crowdsourced monitoring systems where AI enables diversity in citizens contributing on various topics, from pollution and traffic flow to safety, while analyzing resultant data into usable insights for policy actors—the collaborative problem-solving aspect of the work. AI-driven voter assistance tools would further train the electorate on policy and candidates for informed choices, while platforms involved in citizen science marshal the various expertise of different types of people in scientifically studying and monitoring environmental conditions. This is a context that breeds ecology in knowledge sharing and collective responsibility (Lima et al., 2022).

Ethics and transparency AI panels—these avail a discussion platform for the citizenry on the various ethical dimensions of different applications. They serve to help lay down guidelines and accountability frameworks through which confidence in the AI technologies can be instilled. AI Social Media Monitoring Tools – these help identify emerging civic issues, hence enabling governments to respond effectively and timely (Sharma et al., 2021; Bayerl et al., 2023).

2.3 CITIZEN PARTICIPATION THROUGH AI

Citizen participation is a necessary ingredient for a sustainable, effective, inclusive urban environment. Specifically for cities, AI can enhance such participatory processes and

TABLE 2.3 The AI Role in Citizen Participation in Smart Governance

Area of Participation	Impact on Citizen Engagement	Examples of Applications
Urban planning and design	Empowers citizens with information and forecasts on urban projects for better and more inclusive involvement in decision-making processes.	Virtual public consultations using AI simulations of city developments, AI tools for collecting public feedback on urban designs.
Public services management	It improves citizen engagement by making timely information available and providing direct channels to report issues, thereby allowing the services to be responsive and user-centered.	AI chatbots for reporting infrastructure issues like broken streetlights, predictive maintenance for waste management systems.
Traffic and mobility	Engages citizens in the development of the city's transportation system through data on traffic patterns and discussions on solutions for mobility.	AI tools for citizen input on smart parking spaces, AI-based traffic apps offering citizens the ability to influence city traffic flow.
Energy consumption and sustainability	Engages citizens to take more interest in sustainability initiatives, using real-time energy data to help make better choices regarding energy conservation.	AI-powered platforms for monitoring personal energy consumption, AI tools for creating community-wide sustainability programs.
Healthcare and well-being	Involves citizens more in health-related decisions through real-time feedback about the needs within the community for health, creating personalized wellness plans.	AI-driven health monitoring apps for citizens, AI tools for public health surveys to understand community needs.
Citizen feedback systems	Citizen engagement is driven by automatically analyzing huge volumes of public feedback and extracting from it insights that can be put to action by city planners and policymakers.	AI-driven platforms for public feedback on city policies, NLP tools for analyzing citizen complaints and suggestions.
Public safety and security	It enhances civic engagement by enhancing openness, allowing easier interaction between communities and the police, and involves citizens in safety initiatives.	AI tools for community-based crime reporting, predictive policing and emergency alert systems for public safety.
Social inclusion	Inclusivity will be increased through the availability of city services to persons with disabilities and non-native language groups, improving their participation.	AI-based translation services for non-native speakers, AI tools for providing accessibility options in public spaces and services.
Environment monitoring	Environmental factors in real-time will help citizen engagement to encourage joint actions around issues like air pollution, waste management, and green spaces.	AI systems for crowd-sourced environmental monitoring, Smart City dashboards displaying environmental data for public awareness.
Voting and e-governance	Engages more active citizenship in governance, with a secured, efficient, and accessible voting mechanism for increased democracy in urban governance.	AI platforms for electronic voting and civic engagement apps, AI-powered platforms for monitoring e-government interactions.

Source: Author's own work.

enable residents to interact better with the services, policies, and developments in their cities. Table 2.3 presents in detail how AI facilitates citizen participation in Smart Cities.

AI can now provide smart governance through automation and by providing advanced analytics, while in the future directions, the area involves AI-powered decision support systems analyzing huge data volumes, identifying patterns, and presenting data-driven recommendations as shown in Table 2.4. Such systems empower policymakers to create more informed and effective policies based on fact-based insights and predictive analytics (Zhao, 2023). In addition, AI makes it possible to forecast socio-economic trends and events that are likely to give rise to future risks, hence enabling proactive governance and crisis prevention for greater efficiency and resilience in dealing with modern challenges (Jonek-Kowalska and Wolniak, 2021).

Other development has to do with the intelligent automation of public services to increase administrative efficiency. This sort of automation will minimize bureaucratic delays, operational costs apart from document processing, application reviews, and management of citizen data (Antenucci and Meissner, 2024). AI chatbots further smoothen communication with the citizenry in order for them to be more accessible and satisfactory to the citizenry. AI is what enables personalization in public services: through analyzing the preferences and behavior of the users, it comes forth with a solution closer

TABLE 2.4 Future Directions of AI in Smart Governance

Future Direction	Description
AI-driven decision support systems	Use of AI-powered tools for large dataset analysis, pattern identification, and data-driven recommendations to policymakers for better decision-making quality and efficiency.
Predictive analytics for policymaking	It means using predictive models to forecast societal trends, economic changes, and potential crises to enable proactive governance and pre-emptive problem-solving.
Intelligent automation of services	Automate administrative tasks and public services with AI, reducing bureaucratic times, increasing efficiency in service delivery, and raising citizen satisfaction.
Enhanced citizen engagement platforms	The development of AI-enabled platforms to support interactive communication, gather feedback, and conduct sentiment analysis to ensure participatory and inclusive governance.
AI-powered risk management	Application of AI systems in the identification and reduction of risks from security, climate change, and infrastructure for long-term sustainability and resiliency.
Ethical AI frameworks	Formulation of guidelines and frameworks that would make the use of AI ethical in order to overcome biases, data privacy issues, and fairness for accountability in governance.
Smart urban planning	Use AI in real-time monitoring and optimization of urban infrastructure: flow, energy management, waste management, etc., assuring sustainability for urban development.
Fraud detection and transparency tools	Leveraging AI algorithms in detecting anomalies, identifying corruption risks, and fostering transparency in governmental operations and financial transactions.
Personalized public services	Tailing public services to the needs and preferences of citizens analyzed by means of AI guarantees appropriate solutions and improves the accessibility of these services.
AI-based crisis management systems	Development of AI-driven frameworks for disaster response, emergency management, and resource allocation to assure rapid and effective handling in crisis situations.

Source: Author's own work.

to each of their needs. Services will therefore become more inclusive and responsive, becoming adaptive and citizen-centric (Rajagopal et al., 2023).

The next relevant domain in which AI will surely influence smart governance in the future relates to the improved security mechanisms for transparency. Fraud detection algorithms could turn useful for unmasking corruption and the misappropriation of funds with added accountability and equity within public administration. AI makes for improved risk management and crisis response, faster decision-making, and mobilization processes against disasters or cybersecurity attacks (Wolniak and Stecuła, 2024). Where there is a continuous evolution in AI technologies, there exists concurrently a need for the building up of ethical frameworks in order to have responsible uses that are nondiscriminatory. Setting standards will help the public feel more confident in adopting the AI-driven model of governance that is fair and sustainable.

BIBLIOGRAPHY

Al-Besher, A., and Kumar, K. (2022). Use of artificial intelligence to enhance e-government services. *Measurement and Sensors*, 24, 100484.

Antenucci, I., and Meissner, F. (2024). AI and urban governance: From the perils of smart cities to Amazon Inc. urbanism. In: R. Paul, E. Carmel, & J. Cobbe (Eds.), *Handbook on Public Policy and Artificial Intelligence*. Cheltenham, Great Brittan: Edward Elgar Publishing, pp. 423–434.

Arora, A., Jain, A., Yadav, D., Hassija, V., Chamola, V., and Sikdar, B. (2023). Next generation of multi-agent driven smart city applications and research paradigms. *IEEE Open Journal of Communications Society*, 4, 1.

Ashwini, B.P., Savithramma, R.M., and Sumathi, R. (2022). Artificial intelligence in smart city applications: An overview. In *Proceedings of the 2022 6th International Conference on Intelligent Computing and Control Systems (ICICCS)*, Madurai, India, 25–27 May 2022. Piscataway, NJ: IEEE, pp. 986–993.

Barragán Vargas, D.A., Gonzalez Bustamante, R.A., and Escobar, R.F. (2022). Artificial intelligence, case study: Detection of diabetic retinopathy through neuronal networks in citizens of Bogotá, Colombia. In *Proceedings of the International Conference on Knowledge Management in Organizations*, Hagen, Germany, 11–14 July 2022. Berlin/Heidelberg: Springer, pp. 372–392.

Bayerl, P.S., Bates, L., and Akhgar, B. (2023). Securing the smart city: Patterns of public acceptance for integrated technological solutions. In *Proceedings of the 2023 IEEE International Smart Cities Conference (ISC2)*, Bucharest, Romania, 24–27 September 2023. Piscataway, NJ: IEEE, pp. 1–6.

Bokhari, S.A.A., and Myeong, S. (2022). Artificial intelligence-based technological-oriented knowledge management, innovation, and e-service delivery in smart cities: Moderating role of e-governance. *Applied Sciences*, 12, 8732.

Bokhari, S.A.A., and Myeong, S. (2023). The impact of AI applications on smart decision-making in smart cities as mediated by the Internet of Things and smart governance. *IEEE Access*, 11, 120827–120844.

Bosco, G., Riccardi, V., Sciarrone, A., D'Amore, R., and Visvizi, A. (2024). AI-driven innovation in smart city governance: Achieving human-centric and sustainable outcomes. *Transforming Government: People, Process and Policy*, 18(4), 423–434.

Caprotti, F., and Liu, D. (2022). Platform urbanism and the Chinese smart city: The co-production and territorialisation of Hangzhou City Brain. *GeoJournal*, 87, 1559–1573.

Castillo-Villar, R.G. (2021). Contemporary challenges to university governance models. In: M.A. Khan, A.J. Dieck-Assad, R.G. Castillo-Villar, & T.K. Henderson-Torres (Eds.), *Governance Models for Latin American Universities in the 21st Century: Comparative Analysis, Global Perspectives, and Future Propositions*. Cham, Switzerland: Springer, pp. 59–71.

Chiordi, S., Desogus, G., Garau, C., Nesi, P., and Zamperlin, P. (2022). A preliminary survey on smart specialization platforms: Evaluation of European best practices. In *Proceedings of the International Conference on Computational Science and Its Applications*, Malaga, Spain, 4–7 July 2022. Berlin/Heidelberg: Springer, pp. 67–84.

Davide, F. (2021). Perspectives for digital social innovation to reshape the European welfare systems: An introduction. In *Perspectives for Digital Social Innovation to Reshape the European Welfare Systems*. IOS Press, pp. 1–41.

Dong, L., and Liu, Y. (2023). Frontiers of policy and governance research in a smart city and artificial intelligence: An advanced review based on natural language processing. *Frontiers in Sustainable Cities*, 5, 1199041.

Downes, L., Garad, R., Teede, H.J., and Norman, R.J. (2018). A governance model addressing the challenges of contemporary research. *Seminars in Reproductive Medicine*, 36(1), 73–79.

Fabregue, B. (2024). Artificial intelligence governance in smart cities: A European regulatory perspective. *Journal of Autonomous Intelligence*, 7, 1–16.

Guarda, T., Lopes, I., Bustos, S., Ribeiro, I., and Fernandes, A. (2023). Augmented computing and smart cities sustainability. In *Proceedings of the International Conference on Computational Science and Its Applications*, Athens, Greece, 3–6 July 2023. Berlin/Heidelberg: Springer, pp. 123–132.

Harnal, S., Sharma, G., Malik, S., Kaur, G., Khurana, S., Kaur, P., Simaiya, S., and Bagga, D. (2022). Bibliometric mapping of trends, applications and challenges of artificial intelligence in smart cities. *EAI Endorsed Transactions on Scalable Information Systems*, 9, e8.

Jain, S., Kapur, S., and Dobhal, V.C. (2021). Virtually interactive user manual for command and control systems using rule-based chatbot. In *Proceedings of the Artificial Intelligence and Sustainable Computing for Smart City: First International Conference, AIS2C2 2021*, Greater Noida, India, 22–23 March 2021. Berlin/Heidelberg: Springer, pp. 162–172.

Jonek-Kowalska I., and Wolniak R. (2021). Economic opportunities for creating smart cities in Poland: Does wealth matter?. *Cities*, 114, 103222, 1–16.

Jonek-Kowalska, I., and Wolniak, R. (2023). *Towards Sustainability and a Better Quality of Life?* Routledge: London, UK.

Kuguoglu, B.K., van der Voort, H., and Janssen, M. (2021). The giant leap for smart cities: Scaling up smart city artificial intelligence of things (AIOT) initiatives. *Sustainability*, 13, 12295.

Li, W., Yigitcanlar, T., Nili, A., and Browne, W. (2023). Tech giants' responsible innovation and technology strategy: An international policy review. *Smart Cities*, 6, 3454–3492.

Lima, L.B. de, Iano, Y., Noritomi, P.Y., Oliveira, G.G. de, and Vaz, G.C. (2022). Data security, privacy, and regulatory issues: A conceptual approach to digital transformation to smart cities. In *Proceedings of the Brazilian Technology Symposium*, 24–26 October 2022. Berlin/Heidelberg: Springer, pp. 256–263.

Ng, L.H.X., Lim, A.C.M., Lim, A.X.W., and Taeihagh, A. (2023). Digital ethics for biometric applications in a smart city. *Digital Government: Research and Practice*, 4, 1–6.

Oleksy, T., Wnuk, A., Domaradzka, A., and Maison, D. (2023). What shapes our attitudes towards algorithms in urban governance? The role of perceived friendliness and controllability of the city, and human-algorithm cooperation. *Computers in Human Behavior*, 142, 107653.

Pahuja, N. (2022). Partnering with technology firms to train smart city workforces. In *Smart Cities Policies and Financing*. Amsterdam: Elsevier, pp. 169–180.

Pribadi, U., Juhari, Ibrahim, M.A., and Kurniawan, C. (2023). Pivotal factors affecting citizens in using smart government services in Indonesia. In *Proceedings of the International Congress on Information and Communication Technology*, London, UK, 20–23 February 2023. Berlin/Heidelberg: Springer, pp. 1087–1099.

Rajagopal, M., Sivasakthivel, R., Ramar, G., Mansurali, A., and Karuppasamy, S.K. (2023). A conceptual framework for AI governance in public administration – a smart governance perspective. *7th International Conference on I-SMAC* (IoT in Social, Mobile, Analytics and Cloud), I-SMAC 2023 – Proceedings, pp. 488–495.

Saadah, M. (2021). Artificial intelligence for smart governance: Towards Jambi smart city. *IOP Conference Series: Earth and Environmental Science*, 717, 12030.

Saptadi, N., Suyuti, A., Ilham, A.A., and Nurtanio, I. (2023). Literature study on the role of artificial intelligence waste management into biomass briquettes toward smart city governance. *AIP Conference Proceedings*, 2680, 020048.

Schintler, L.A., and McNeely, C.L. (2022). Artificial intelligence, institutions, and resilience: Prospects and provocations for cities. *Journal of Urban Management*, 11, 256–268.

Sharma, S., Gianoli, A., and Geerlings, H. (2021). Governance of ambient assisted living for age-friendly inclusive cities. In *Proceedings of the 3rd International Conference on Human Interaction and Emerging Technologies: Future Applications (IHIET 2020)*, Paris, France, 27–29 August 2020. Berlin/Heidelberg: Springer, pp. 612–617.

Sienkiewicz-Małyjurek, K. (2023). Whether AI adoption challenges matter for public managers? The case of Polish cities. *Government Information Quarterly*, 40, 101828.

Singh, K.D., Singh, P., Chhabra, R., Kaur, G., Bansal, A., and Tripathi, V. (2023). Cyber-physical systems for smart city applications: A comparative study. In *Proceedings of the 2023 International Conference on Advancement in Computation & Computer Technologies (InCACCT)*, Gharuan, India, 5–6 May 2023. Piscataway, NJ: IEEE, pp. 871–876.

Skubis, I., Wolniak, R., and Grebski, W.W. (2024). AI and human-centric approach in smart cities management: Case studies from Silesian and lesser Poland voivodships. *Sustainability*, 16(18), 8279.

Stecuła, K., Wolniak, R., and Aydın, B. (2024). Technology development in online grocery shopping—from shopping services to virtual reality, metaverse, and smart devices: A review. *Foods*, 13(23), 3959.

Stecuła, K., Wolniak, R., and Grebski, W.W. (2023). AI-driven urban energy solutions—from individuals to society: A review. *Energies*, 16, 7988.

Ullah, Z., Naeem, M., Coronato, A., Ribino, P., and De Pietro, G. (2023). Blockchain applications in sustainable smart cities. *Sustainable Cities and Society*, 97, 104697.

Vrabie, C. (2022). Artificial intelligence promises to public organizations and smart cities. In *PLAIS EuroSymposium on Digital Transformation*. Berlin/Heidelberg: Springer, pp. 3–14.

Wolniak, R., and Stecuła, K. (2024). Artificial Intelligence in smart cities—applications, barriers and future directions: A review. *Smart Cities*, 7, 1346–1389.

Wu, B. (2022). Research on the application of computer artificial intelligence technology in the digitalization of modern cities. In *Proceedings of the 2022 IEEE Conference on*

Telecommunications, Optics and Computer Science (TOCS), Dalian, China, 1–12 December 2022. Piscataway, NJ: IEEE, pp. 272–276.

Xiao, X., Xie, C., and Thomson, G.W. (2021). Research on the application of artificial intelligence in intelligent city and its countermeasures. *Journal of Physics: Conference Series*, 1910, 12015.

Zhao, H. (2023). Artificial intelligence-based public safety data resource management in smart cities. *Open Computer Science*, 13, 20220271.

Zhou, J. (2022). Artificial intelligence-based recommendation and application of public services in smart cities. *Computational Intelligence and Neuroscience*, 2022, 8958865.

CHAPTER 3

AI in Smart Economy

3.1 AI-DRIVEN ECONOMIC TRANSFORMATION

The Artificial Intelligence (AI)-driven economic transformation of a Smart City presents a mutating moment in how an urban setting works and how economic values arise. Core to that, AI technologies promise huge opportunities for reshaping urban life in Smart Cities by effective use of resources, higher productivity, and unprecedented improvements in lifestyles of its citizens. It's not a tech upgrade alone; rather, it is holistic remodeling in every scale of city performance by reimagining infrastructure, governance, public services, mobilities, and economic systems to use large data and automation opportunities (Dawar and Kumar, 2024).

AI may also have implications for the economy of the cities when trying to optimize operations and use of resources within Smart Cities. These range from smart systems operating utilities efficiently, consumption of reduced energy, to effective handling of waste (Stecuła et al., 2023). They use predictive analytics to estimate demand, flow in traffic, and ensure delivery at the lowest cost for fundamental services such as water, electricity, and transportation. Their contribution, besides reducing operation costs, is toward sustainability in that cities are able to sustain growing populations without complete draining of natural resources (Cirella et al., 2023; Jonek-Kowalska and Wolniak, 2023).

Beyond the operation efficiencies, AI is making it possible for businesses and industries to interface with one another across Smart Cities (Wolniak et al., 2024). The rise of IoT means more devices and machines in this interrelated ecosystem represent not just devices and machines that talk to one another and disseminate data within an analytic system to make economic sense, but also much else. AI-driven insight helps businesses develop products and services that are in line with the needs of citizens, increasing their

DOI: 10.1201/9781003587613-4

satisfaction, which helps grow the local economy (Wolniak and Jonek-Kowalska, 2022; Wolniak and Stecuła, 2024). Apart from all these, AI creates new business models, particularly in those industries where automation and data-driven solutions are finding more traction, including logistics and transportation, healthcare, and retail (Ray et al., 2023; Xu, 2022).

Probably the most important consequence that could be coming for Smart Cities within the labor market is a shift in the economy toward AI-driven endeavors. While AI might innovate and develop in fields yet unknown, therefore opening up other opportunities that may employ workers, one factor that has been necessary, so far, has been restructuring in the available worker skill sets. Correspondingly, rapid increase in demand is manifest in the availability of skilled manpower capable of designing, installing, and managing such systems. This has triggered off a number of trainings and educative initiatives with the aim of making people capable of succeeding within the framework of the technology-intensive economy. On the other hand, fears related to loss of jobs will arise since routines, whether of a manual or administrative nature, will be the most vulnerable. The question then becomes one of social policies that tend to address adaptation to the new economy of the workforce and the equitable distribution of benefits accruing from economic growth. The transformation of the economy with the aid of AI has also created changes in how cities interact with their citizens.

Diffused in all public services, such as healthcare, education, and welfare, AI is able to render these services more effective and accessible. For example, AI can analyze health data to predict outbreaks, thus making personalized medical treatments possible; this would make the health system more responsive to people's needs. Also in education, AI can more individually tailor learning experiences to suit the needs and abilities of each student, improving outcomes and better preparing future generations for the evolving demands on the labor market (Wang et al., 2023; Bhukya et al., 2023).

It will also contribute to growth and the creation of smart infrastructures in cities. AI can even enable much better urban growth forecasting, superior land use, and superior management of public spaces through intelligent urban planning. By incorporating AI, a city can turn responsive to continuous changes within environmental conditions, demographic trends, and technological development. Such a proactive stance henceforth enables the Smart Cities to handle not only the problems arising in the future but also such needs, thus making it resilient and sustainable (Bhukya et al., 2023).

This influence of AI, being performed at a Smart City level, finds its reflection even on the global level. While the towns become more smart and effective, their competitiveness increases internationally. They improve living with integrated AI, attract investment, innovate, and grow economically. Companies and startups will become more and more attracted to this high-tech urban area as such infrastructure allows them to have fast development and implementation of solutions with AI in place. It thus plays a major role in wider economic growth of the nations where cities having AI-driven transformation become economic hubs for different activities and innovations (Sikos Tomay and Szendi, 2023; Elahi et al., 2021).

AI can remake industries through raising productivity and competitiveness and lowering costs. More generally, the general diffusion of the technology could result in an efficient, dynamic, innovative global economy, with spillovers from AI-driven growth spread across many industries, offering new opportunities and propelling sustainable development (Table 3.1).

3.2 OPPORTUNITIES AND CHALLENGES IN THE AI ECONOMY

High-fast-growing populations, increased scarcity of resources, and mounting concerns for the environment have brought the need for smart urban areas to the fore. AI has perhaps emerged as the most powerful tool really to shape the urban landscape today. It symbolizes this visionary dream where the city can be more effective, more sustainable, and more "livable"—Smart Cities with AI empowerment. This transformation is also not only in integrating new technologies but also in redefining how cities operate—economically, socially, and environmentally (Zhou, 2022). The potential of AI to alter the economy of Smart Cities lies in its better resource management and increased efficiency in services delivery. AI-powered systems upgrade transportation networks through the use of predictive analytics that reduce congestion, improve public transit routes, and optimize logistics. It is these improvements that have made citizens' lives easier and have contributed to the productivity of the economy by reducing time and cost for transportation. Application of AI in managing traffic—for example, smart traffic light and autonomous cars—can further reduce the amount of emission into the air and make the place much more sustainable (Duan et al., 2022).

AI platforms make cities inclusive; they equip cities to create equity by improving urban planning and promoting economic development in less privileged parts of towns and cities. Such patterns of inequity, through data analysis done by AI tools, support informed decisions to make strategic investments in infrastructure, education, and public services. Automation of the administrative tasks decreases the operation cost for running an administration and increases governance, thereby making governments efficient, transparent, and responsive to citizens. It also contributes to the betterment of the business environment, enables innovation, and provides benefits to the economy through the involvement and participation of each and every citizen (Streitz et al., 2022).

Smart Cities driven by AI open a new economic sector, too. AI technology developed, tested, and implemented in the urban environment feeds into growth in robotics, data science, and cybersecurity among others, aside from smart infrastructure. This includes a wave of startups, with established ones, too, venturing into Smart City-specific solutions, which ascertains a thriving ecosystem for innovations and job opportunities. It will allow AI integration into industries all the way from real estate and construction to finance and education. It will also enable new business models based on collaboration (Sarjana, 2023).

AI, while integrated into the development and functioning of Smart Cities, contributes significantly to the avenues of economic transformation. Table 3.2 presents some key AI-driven opportunities, along with how exactly each of these opportunities represents a way toward economic development in an urban setting.

TABLE 3.1 The Role of AI in Transforming Industries and Driving Economic Growth

Industry	AI Applications and Transformations	Potential Benefits
Manufacturing (Yu et al., 2024; Kiefer et al., 2024; Gavade, 2023)	Automation and Robotics: AI-driven robots and automation systems ease the production process, reducing or eliminating human fallacies. Predictive Maintenance: AI makes failure estimates on equipment before actual occurrences by analyzing the data that is generated by them, hence reducing downtimes. Supply Chain Optimization: AI helps in demand forecast optimization, optimization of logistics, and enhancement in inventory management. Quality Control: Through the use of machine learning, the defects or abnormalities which may be manufactured in the product during its production will get picked up for improvement.	Increased Productivity: The system automates repetitive tasks, and workers can devote more time to higher-value activities. Cost Reduction: With predictive maintenance and optimized supply chains, operational costs will be reduced. Improved Competitiveness: With AI-driven demand forecasting and supply chain management, manufacturers would be able to respond promptly to market changes.
Finance (Reshmi et al., 2024; Kuang et al., 2024)	Algorithmic trading: AI goes through Big Data to make high-frequency trading decisions; therefore, optimization of investment strategy is enhanced. Risk Assessment and Fraud Detection: The machine learning algorithms analyze transactional data for pattern detection in fraud activities and assess the risks. AI-driven Virtual Assistance and Chatbots provide advice on financial matters with personalized experience and 24*7 assistance. Portfolio Management: With the use of AI to continually analyze real-time data and come up with strategies, thus assisting clients in managing their investments.	Improved Efficiency: Automation of trading and customer service processes drives faster decision-making and cost savings. Lower Risk: Advanced AI models detect fraud, assess financial risks with much greater precision. Competitive Advantage: Financial institutions applying AI can offer more personalized services, thus assuring higher customer loyalty and attracting new clients in their direction.
Healthcare (Kostadinov et al., 2025; Quaranta et al., 2025)	Diagnostics and Imagining: Artificial Intelligence allows evaluating medical images, such as X-rays or MRI scans, for diagnosing diseases like cancers beyond human clinicians. AI in Personalized Medicine: One may permit access to his genetic information in the interest of serving his best interests for an optimal patient outcome. Drug Development: It also speeds up the discovery of new drugs by simulation and analysis of interaction between molecules. Clinical Decision Support: AI provides evidence-based recommendations to healthcare professionals.	Better Patient Outcomes: AI-driven diagnosis, personalized treatments—translation to highly accurate diagnosis and better treatment plans. Cost Savings: Automation of administrative tasks and predictive analytics brings down the cost of healthcare. Increased Efficiency: AI accelerates drug development and clinical decision-making in drug development, accelerating the time to market for new treatments.
Retail (Lopes et al., 2025; Shanti, 2025; Petrescu et al., 2024; Wolniak et al., 2024; Stecuła et al., 2024)	Personalized Recommendations: Through customer behavior and preference analysis, AI makes personalized recommendations of products. Inventory and Supply-Chain Management: AI supports demand forecasting to maintain an optimum level of stock and avoid overstocking or stockouts. Improved Customer Experience: AI-driven chatbots set a new level of expectations for customer experience. Price Optimization: The AI algorithms run dynamic price adjustments by calculating demand, competition, and market conditions.	More Sales and Customer Retention: Personalization makes customers happy, therefore loyal, frequently returning. Operational Efficiency: Optimal inventory and pricing strategy leads to less waste and better margin. Competitive Advantage: AI-driven insight into consumer behavior is of essence to retailers for competitive advantage in responding to trends in the marketplace and consumer needs.

(*Continued*)

TABLE 3.1 (Continued)

Industry	AI Applications and Transformations	Potential Benefits
Transportation (Pang et al., 2024; Huszák et al., 2024; Lukic Vujadinovic et al., 2024)	Auto Drive: With the power of AI, cars and trucks can self-drive. This reduces human error, and traffic flow can easily be optimized. Route Optimization: AI algorithms seek the best route to take. This reduces time and fuel consumption in the field of logistics. Fleet Management: Using machine learning on vehicle performance can help analyze data regarding scheduling maintenance and monitoring drivers' behavior. Smart Cities and Infrastructure: AI in intelligent traffic management helps avoid traffic congestion by optimizing flows of urban transportation.	Reduced Costs: AI-driven route optimization and autonomous vehicles lower fuel consumption and labor costs. Increased Safety: Self-driving vehicles and AI-powered traffic management reduce the chances of accidents. Environmental Benefits: Optimized transportation networks result in lesser emissions and a reduced carbon footprint.

Source: Author's own work.

Most AI technologies are capital-intensive in the collection of data systems, high-speed connectivity, and cloud computing platforms. This requires large volumes of data that AI systems have to acquire, process, and analyze in real time—a task that a well-developed digital infrastructure does very well, which may presently not be available to most cities, especially those in developing regions. In better-developed cities, upgrading obsolete systems to a level capable of hosting such AI technologies is extremely expensive and labor-intensive, hence beyond the reach of many ambitious projects taken by local governments. In very many cases, the economic transformation brought by AI presupposes complex systems of algorithms and data processing—not only very expensive but also with high demand for specific human resources. Indeed, highly talented experts in data science, engineering, and AI expertise are still highly sought after, and their possible shortage could limit to a large extent the ability of Smart Cities to apply AI effectively. The scarcity is further worsened in that the competition for skilled workers in this field is amazingly high. Because AI is powered by machine learning algorithms, success with any AI system totally depends on what is fed into the system in terms of data. Partial, biased, or incomplete data will skew the results toward ineffective or even dangerous policies that further complicate the challenges cities face in applying AI (Kalenyuk et al., 2023; Tamás and Dóra, 2023).

Other challenges include the ethical and social impact of AI. While Smart Cities use AI to manage services and infrastructure, this raises ethical and security concerns (Table 3.3). The vast information that will be collected by AI systems, even on something as personal as each citizen, presents a great chance for privacy breach. The use of AI in most situations will be regarded as intrusion into the lives of citizens and, as such, is going to further evoke the mistrust of the general public. Other ethical issues concerning AI, which still need discussing, are that such algorithms generate prejudicial decisions for which an AI system can serve or sustain discrimination. That is a lot of thinking, clear regulatory frameworks, and continuous oversight that have to go into making the AI systems work transparently and in an ethical manner, serving all citizens (Alaeddini et al., 2023).

TABLE 3.2 Opportunities of AI-Driven Economic Transformation of Smart Cities

Opportunity	Description
Optimization of urban infrastructure	This would, in turn, improve most of the infrastructural systems, particularly those related to energy, water, and waste management in a city. Application of AI-based predictive maintenance in the city would mean an effective affordable way of enhancing efficiency with asset life extension; hence, sustainable and resource-efficient urban areas invite businesses to make lives pleasant.
Smart mobility and transportation	AI technologies in Smart Cities can facilitate the creation of efficient, ecological, and even safer transport systems. Traffic flow would be optimized, reducing congestion with the use of AI and making the deployment of autonomous vehicles possible. This reduces fuel consumption and transportation cost, hence lowering pollution while increasing economic growth and sustaining the environment. Besides, better mobility could mean productivity from both workers and businesses.
Data-driven urban planning	AI-enabled analytics allow planners to make more insightful decisions from sources such as IoT sensors, social media, and urban scenario simulations. It is also an economic interest in the making of urban spaces more resourceful and attentive, developing better real estate and building innovation ecosystems within city confines. Data-informed city planning will enable them to strategize resources that are potential in bringing overall economic improvements.
Automation of public services	AI will smooth and automate much of the public services, ranging from policing to healthcare to refuse collection. Automation enhances service delivery at lower costs with increased efficiency by organizations in the public sector. This improves living standards for its citizens and operational efficiency in the city—very important factors in economic competitiveness and appeal to businesses.
AI-enhanced healthcare systems	Artificial Intelligence integrated with health services in the Smart City will mean better diagnostics, more personalized treatments, and optimum management of hospitals. In a position to design such predictive tools, AI introduces them with the ultimate aim of reducing medicine costs and improving treatment outcomes. Improved health contributes to economic growth and innovation in new health technologies, creating markets for such technologies, which also attracts doctors and researchers.
Energy efficiency and sustainability	This would, in turn, help AI further enhance the rate of energy consumption and increasingly depend on renewable sources. On the other hand, AI-driven smart grids would be able to monitor and manage energy distribution in real time, reduce waste, and ultimately cut energy costs to businesses and residents. Energy: The sustainable solution contributes to economic development through assuring energy security depending on non-renewable resources and opening up green economic opportunities.
Intelligent resource allocation	AI could also be applied in managing city resources: managing the funds, dispensing resources in cases of crisis, routing of public transport whereby such resources are bound to be used in a much better manner. In all, this will mean much better economic returns with waste reduced and efficiency increased in most industries, thus being more attractive to investors and entrepreneurs.
Creation of high-tech jobs and skills	Artificial inclusions within the urban economy give rise to new industries, businesses, and therefore jobs: Data Science, Development, and Cybersecurity. This helps increase education and skilled-based training, hence is complementary to this job in developing large-scale economic development on the strong resource of manpower.
AI-driven public safety and security	Facial recognition has many great things it does: predictive policing, surveillance, a host of other AI-powered technologies go a long way in the improvement of public safety, reduction of crime rates, and safer urban environments. With safe and secure cities, coming investors, residents, and businesses are guaranteed; hence, it is a guarantee of overall economic stability and growth for the region. Safety guarantees tourism and business, which must be a big part of the local economy.

Source: Author's own work.

TABLE 3.3 Challenges of AI-Driven Economic Transportation of Smart Cities

Challenge	Description
Data privacy and security concerns	AI systems process a quantity of sensitive personal and business data, generating again risks about privacy, data breaches, and misuse. Ensuring strong data protection mechanisms and earning public trust is a big challenge. While there is a lack of security protocols, AI can let down citizens and businesses while trading off economic benefits.
High initial investment costs	AI adoption in Smart Cities has huge upfront investments in infrastructures, technologies, and human resources. The creation cost of systems for AI-powered mechanisms is also quite high for any developing city or a government with a limited economy to afford, which leads to delays or barriers toward full-scale AI implementation within a Smart City initiative.
Technological unemployment	Automation of services like public transport, health, and police would result in the loss of many jobs. Substitution of jobs that earlier used to exist may get displaced by AI technologies, which in turn would result in unemployment or underemployment in low-skilled sections of the population and bring forth social inequity with economic instability.
Digital divide and inequality	These inequalities in the availability of AI technologies will surely exacerbate digital gaps within disprivileged groups. Once again, not all of them enjoy equally available high-speed access to the Internet inside the Smart City—neither are other sophisticated technologies nor even other programs for digital literacy equitably distributed across all citizens and businesses. The situation could thus aggravate social disparities, with the potential even making it harder to attain economic progress toward these disadvantaged communities overcoming their long-standing socio-economic challenges.
Regulatory and ethical challenges	With an increased integration of AI into the urban system, most specifically, the question arises of whether such innovation is adequately regulated. Policymaking, which must be fair and effective, becomes an increasingly impossible process as the technologies of AI are still in the middle of a rapidly changing process. Other possible pitfalls may include ethical ones, for instance, biased algorithms or results. It is up to governments and policymakers to make sure this adoption is nondiscriminatory, fair, and transparent.
Lack of standardization and interoperability	AI technologies being put to work in Smart Cities stem from many vendors and various platforms that, of course, are not standardized and non-interoperable. It creates silos whereby several AI applications face a lot of hassle to inter-communicate or work with each other. Due to such difficulties, the effectiveness of the solution provided by AI itself gets undermined, hence fragmented and less economically effective service delivery.
Resistance to change	Yet, much resistance against the diffusion of those AI technologies aligns with citizens or enterprises' fear because of its effect on jobs and processes in control, or just basically strange technology. Situations that have these resistances are in need of educational activities, awareness, and indication that AI would add value in a longer-term perspective, cases in which it is obtained gradually.
High complexity of AI integration	Integrate the AI technologies into the prevailing urban structure that is essentially technical and coordination challenges among various sectors. In such context, seamless integration of AI-driven transportation, energy, and public service delivery mechanisms can be effectively merged with Smart City systems. The presence of such challenges can easily introduce complexity and further delays with cost overruns, therefore hindering full-scale implementation.
Maintaining human oversight	Accountability, fairness, and transparency—all these need human checks, even though much of the functions can now be performed automatically by AI. Too much dependency upon AI might lead to undesirable outcomes such as errors in algorithms or ethical lapse. Balance between autonomy provided to AI and its checking by humans will have to be drawn appropriately so that adverse impacts on economic and social aspects do not occur.
Environmental impact of technology	Full-scale use of AI, along with related technologies at data centers, sensors, and connectivity infrastructure, requires huge amounts of energy; this highly contributes to the degradation of the environment. The design of energy-efficient and sustainable AI systems balances economic transformation with environmental responsibilities.

Source: Author's own work.

Governance and regulation— AI technologies form part of Smart City functionalities, policy development by governments needs to make sure that such technologies are responsibly and effectively deployed—in other words, setting up rules that will protect security and privacy without quelling innovation, which is no easy fit. In fact, the governance of AI can easily become an international affair since virtually all cities and countries are no longer discrete entities. The call is for global standards and frameworks to manage the rollout of AI, and such negotiations may be problematic since different stakeholders have different priorities and resources (Darmawan et al., 2023).

First are economic consequences of AI-driven change that have to be considered in a very reflective way. As much as AI might promise some unprecedented economic growth because of the efficiency improvements it causes in many sectors, it also presents possibilities that the same causes job displacement. Large groups of the working population might find their skills no longer in demand inasmuch as AI systems take up automation in manufacturing, logistics, and service industries. It may spur unemployment rates and further social inequality if no measures to retrain workers and give them new jobs related to AI are taken. There is also a risk of unequal benefits accruing to AI, since the more prosperous regions or companies would be in a better position to exploit these technologies at others' expense.

These challenges need responses at many levels. It is now time for the cities to buckle down and develop robust digital infrastructure, equipping themselves with the necessary toolset and resources to implement AI solutions. Investment in education and training programs is very important to upskill the workers and produce the required talent to manage and maintain AI technologies. While there can be certain inhibitions among educational institutions, partnerships—particularly between these sectors and international organizations—can also be very instrumental in introducing innovation and knowledge sharing, and hence make the development environment for AI more collaborative (Jonek-Kowalska and Wolniak, 2022).

This AI-driven transformation of the economy will indeed have to be fully inclusive. In this direction, policymakers will have to frame policies for workers who are deprived of jobs due to automation and support them with programs involving reskilling, social safety nets, and employment creation in new industries. Indeed, public-private partnership in smoothly facilitating transition in an AI-driven economy ensures that benefits accruing from these technologies must equitably reach all segments of society.

3.3 ADDRESSING WORKFORCE DISPLACEMENT AND SKILL DEVELOPMENT

High-tech integration, especially in the use of AI within the paradigm of a smart economy for a Smart City, is bound to make a sea-change in the very character of the workmanship. Moving cities, or, in other words, cities that fully implement digital infrastructures with AI-driven solutions, also raise an alarm for comprehensive skill development strategies with their growing concern for workers' displacement. The very same change promises enormous economic gain and improvements in livelihood in towns and cities. However, it is equally confronted with critical challenges, most especially in the area of job employment (Lamnatou et al., 2024).

Workforce replacement is a major challenge in Smart Cities. Most industries run a high risk of incurring huge losses daily with the increasing AI and automation technologies doing tasks that were traditionally handled by humans. Such industries include manufacturing, transportation, and administrative services. While industries in logistics and transportation are threatened by driverless cars, AI-based software substitutes for human resources in areas of customer service, data entry, and so on. The outcome of such substitutions is not limited to unskilled jobs alone; even skilled ones are victimized in the process—right from decision-making and management levels, where the tools of AI have begun to support long-range predictive analytics, even strategy. This is further exacerbated by the pace at which those technologies are developing, which simultaneously means most employees will be without the skills and competencies needed to transfer into other professions in a substantially automation-dominated economy (Ojha and Nikhil, 2024).

The challenges provide a test of thinking-ahead art for policymakers and business leaders in forging strategies balancing technological advancement with human workforce sustainability. Skill development has become crucial. What is urgently needed is collaboration between governments and companies in putting together training programs that could prepare workers for a smart economy, not only by equipping them with just the technical skills needed to thrive in this new economy in areas such as data science, machine learning, and programming but also giving them the soft skills comprising creativity, emotional intelligence, and critical thinking. While AI systems are not likely to replace human judgment entirely, there is still high demand for innately human skills: leadership, problem-solving, and collaboration (Skubis et al., 2024).

Educational systems need to change along with technological development. There is an ever-growing need for a new curriculum that would involve early education and vocational training inclusive of AI, digital literacy, and emerging technologies. The culture of lifelong learning will eventually enable workers to adapt to the ever-changing job market. All initiatives undertaken must be inclusive in reaching out to every worker from the formal and informal sectors, as well as every walk of life, so they can retrain. These include some of the most vulnerable populations who are most likely to be displaced, and as such, should be a priority for such initiatives. These initiatives may also include lower-income or less technologically advanced regions (Wolniak and Stecuła, 2024).

Another important agent to facilitate adaptation in the workforce is the role of business leaders. Business leaders must be more forward-looking, not only in the adoption of AI technologies but also in investing in the growth and development of employees. The organization should, besides, have its reskilling programs in-house, create new kinds of jobs where human-AI collaboration will be possible, and embed innovation within its culture. Companies would do well to further study the new models, services, and possibilities thrown into view by AI-enabled opportunities for completely new job segments. This is where, at the policy level, social safety nets and transition mechanisms should exist to take care of workers who are being taken out of their jobs: unemployment benefits, income support, transition programs, financial and educational assistance, among others. They also have to make sure that the adoption of AI technologies does not work in conflict with social equity but actually guarantees a wide diffusion of benefits brought about by the smart economy and keeps it from remaining in a few hands (Jogarao et al., 2024; Akhtar et al., 2024).

A carefully weighted approach to the incidence of AI in workforce displacement and skill development in a Smart City therefore becomes cardinal. Basically, the impact of AI growth would vary greatly with the policies being churned out from the public and private sectors so long as AI growth evolves. It is with preparedness in the workforce, investments in skill development, and social inclusiveness that a city will assure itself its transition into a smart economy does not come at a price taken from the very same individuals powering that very economy. Alternatively said, with the proper precautions in place, AI could be of use to augment human capability, a future where technological extension means the extension of man.

BIBLIOGRAPHY

Akhtar, P., Ghouri, A.M., Ashraf, A., Khan, N.R., and Shuang, M.A. (2024). Smart product platforming powered by AI and generative AI: Personalization for the circular economy. *International Journal of Production Economics*, 273, 109283.

Alaeddini, M., Hajizadeh, M., and Reaidy, P. (2023). A bibliometric analysis of research on the convergence of artificial intelligence and blockchain in smart cities. *Smart Cities*, 6, 764–795.

Bhukya, C.R., Thakur, P., Mudhivarthi, B.R., and Singh, G. (2023). Cybersecurity in internet of medical vehicles: State-of-the-art analysis, research challenges and future perspectives. *Sensors*, 23, 8107.

Cirella, G.T., Domańska, A., and Orobello, C. (2023). Creating smart cities in Poland: Opportunities, obstacles, and the missing link. In: *Proceedings of the International Conference on Smart Technologies in Urban Engineering*, 8–10 June 2023. Springer: Berlin/Heidelberg, Germany, pp. 14–25.

Darmawan, A.K., Muhsi, M., Anekawati, A., Sakdiyah, H., Yusuf, M., Sophan, M.K., Ferdiansyah, D., Umam, B.A., and Jalil, D.K.A. (2023). An interpretive structural model approach to strategic management modelling for sustainable smart village development in Indonesia. In: *Proceedings of the 2023 10th International Conference on ICT for Smart Society (ICISS)*, Bandung, Indonesia, 6–7 September 2023. IEEE: Piscataway, NJ, USA, pp. 1–7.

Dawar, I., and Kumar, N. (2024). Making smart cities smarter: Role of AI in smart cities application. In: H.K.R.V. Rodriguez, Z.Z. Paprika, & A. Ade-Ibijola (Eds.), *Exploring Ethical Dimensions of Environmental Sustainability and Use of AI*. IGI Global: Hershey, PA, USA, pp. 240–262.

Duan, Y., Xu, C., Hu, W., and Zhao, W. (2022). Research on the regulation of artificial intelligence on the orderly development of China's intelligent economy. In: *Proceedings of the 2022 Asia-Pacific Computer Technologies Conference (APCT)*, Wuhan, China, 7–9 January 2022. IEEE: Piscataway, NJ, USA, pp. 11–15.

Elahi, H., Castiglione, A., Wang, G., and Geman, O. (2021). A human-centered artificial intelligence approach for privacy protection of elderly app users in smart cities. *Neurocomputing*, 444, 189–202.

Gavade, D. (2023). AI-driven process automation in manufacturing business administration: Efficiency and cost-efficiency analysis. *IET Conference Proceedings*, 2023(44), 677–684.

Huszák, Á., Simon, V., Bokor, L., Tizedes, L., and Pekar, A. (2024). An AI-driven intelligent transportation system: Functional architecture and implementation. *Infocommunications Journal*, 16(3), 18–30.

Irfan, M., Elmogy, M., Gupta, S., Khalifa, F., and Dias, R.T. (2024). AI-driven decentralized finance and the future of finance. *AI-Driven Decentralized Finance and the Future of Finance*, pp. 1–426.

Jogarao, M., Lakshmanna, B.C., and Naidu, S.T. (2024). AI-enabled circular economy management for sustainable smart cities: Integrating artificial intelligence in resource optimization and waste reduction. *Smart Cities and Circular Economy: The Future of Sustainable Urban Development*, pp. 83–96.

Jonek-Kowalska, I., and Wolniak, R. (2022). Sharing economies' initiatives in municipal authorities' perspective: Research evidence from Poland in the context of smart cities' development. *Sustainability*, 14, 2064.

Jonek-Kowalska, I., and Wolniak, R. (2023). *Towards Sustainability and a Better Quality of Life?* Routledge: London, UK.

Kalenyuk, I., Lukyanenko, L., Tsymbal, L., Stankevics, A., and Uninets, I. (2023). The smart manufacturing: Imperatives and trends. *Financial and Credit Activity: Problems of Theory and Practice*, 5, 327–340.

Kiefer, D., Grimm, F., Straub, T., Bitsch, G., and Van Dinther, C. (2024). Enhancing power skiving tool longevity: The synergy of AI and robotics in manufacturing automation. *International Journal of Mechatronics and Manufacturing Systems*, 17(2), 201–224.

Kostadinov, R., Topalov, V., Georgieva, M., Georgiev, S., and Madzharov, Y. (2025). AI in healthcare—the pre-hospital medical specialists' perception. *Lecture Notes in Networks and Systems*, 883 LNNS, pp. 391–399.

Kuang, M., Kuang, D., Rasool, Z., Saleem, H.M.N., and Ullah, M.I. (2024). From bytes to sustainability: Asymmetric nexus between industrial artificial intelligence and green finance in advanced industrial AI nations. *Borsa Istanbul Review*, 24(5), 886–897.

Lamnatou, C., Cristofari, C., and Chemisana, D. (2024). Artificial Intelligence (AI) in relation to environmental life-cycle assessment, photovoltaics, smart grids and small-island economies. *Sustainable Energy Technologies and Assessments*, 71, 104005.

Lopes, J.M., Gomes, S., Nogueira, E., and Trancoso, T. (2025). AI's invisible touch: How effortless browsing shapes customer perception, experience and engagement in online retail. *Cogent Business and Management*, 12(1), 2440628.

Lukic Vujadinovic, V., Damnjanovic, A., Cakic, A., Vranjes, D., and Bodolo, I. (2024). AI-driven approach for enhancing sustainability in urban public transportation. *Sustainability*, 16(17), 7763.

Ojha, N., and Nikhil, V.P. (2024). Challenges of AI implementation for boosting circular economy in smart cities. In: V. Kandpal, E.D.R. Santibañez-Gonzalez, & P. Chatterjee (Eds.), *Smart Cities and Circular Economy: The Future of Sustainable Urban Development*. Leeds, United Kingdom: Emerald Publishing Limited, pp. 215–233.

Pang, Y., Zhu, X., He, T., Yi, P., and Lee, C. (2024). AI-assisted self-powered vehicle-road integrated electronics for intelligent transportation collaborative perception. *Advanced Materials*, 36(36), 2404763.

Petrescu, M., Krishen, A.S., Gironda, J.T., and Fergurson, J.R. (2024). Exploring AI technology and consumer behavior in retail interactions. *Journal of Consumer Behaviour*, 23(6), 3132–3151.

Quaranta, M., Amantea, I.A., and Molinari, M. (2025). The introduction of AI in healthcare: A multi-layered issue. *Lecture Notes in Computer Science (including subseries Lecture Notes in Artificial Intelligence and Lecture Notes in Bioinformatics)*, 14967 LNAI, pp. 233–244.

Ray, J.K., Sultana, R., Bera, R., Sil, S., and Alfred, Q.M. (2023). A comprehensive review on artificial intelligence (AI) and robotic process automation (RPA) for the development of smart cities. In: S. Bhattacharyya, J.S. Banerjee, & D. De (Eds.), *Confluence of Artificial Intelligence and Robotic Process Automation*. Cham, Switzerland: Springer Nature, pp. 289–311, first online.

Reshmi, L.B., Raj, R.V., Balasubramaniam, S., and Kumar, K.S. (2024). Generative AI and LLM: Case study in finance. In: R.K. Dhanaraj (Ed.), *Generative AI and LLMs: Natural Language Processing and Generative Adversarial Networks*. Berlin, Germany / Boston, USA: De Gruyter, pp. 231–252.

Sarjana, S. (2023). Smart city in supporting sustainable cities. In: *Proceedings of the 2023 10th International Conference on Information Technology, Computer, and Electrical Engineering (ICITACEE)*, Semarang, Indonesia, 31 August–1 September 2023. IEEE: Piscataway, NJ, USA, pp. 365–370.

Shanti, J. (2025). Impact of generative AI on customer experience and personalization in retail sector. *Minds Unveiled: Exploring the Effects of Generative AI on Business Behavior*. London, United Kingdom: Taylor & Francis, pp. 188–196.

Sikos Tomay, T., and Szendi, D. (2023). Analysing economic and environmental sustainability in Hungary. *Urbani Izziv*, 34, 87–97.

Skubis, I., Wolniak, R., and Grebski, W.W. (2024). AI and human-centric approach in smart cities management: Case studies from Silesian and lesser Poland voivodships. *Sustainability* 16(18), 8279.

Stecuła, K., Wolniak, R., and Aydın, B. (2024). Technology development in online grocery shopping—from shopping services to virtual reality, metaverse, and smart devices: A review. *Foods*, 13(23), 3959.

Stecuła, K., Wolniak, R., and Grebski, W.W. (2023). AI-driven urban energy solutions—from individuals to society: A review. *Energies*, 16, 7988.

Streitz, N.A., Riedmann-Streitz, C., and Quintal, L. (2022). From 'smart-only' island towards lighthouse of research and innovation. In: *Proceedings of the International Conference on Human-Computer Interaction*, New Orleans, LA, USA, 30 April–5 May 2022. Springer: Berlin/Heidelberg, Germany, pp. 105–126.

Tamás, S.T., and Dóra, S. (2023). Measuring the economic and environmental sustainability of cities with county rank, 2020–2021. *Teruleti Statisztika*, 63, 89–124. https://doi.org/10.15196/TS630104

Wang, S., Wang, S., Liu, Z., and Zhang, Q. (2023). A role distinguishing Bert model for medical dialogue system in sustainable smart city. *Sustainable Energy Technologies and Assessments*, 55, 102896.

Wolniak, R., and Jonek-Kowalska, I. (2022). The creative services sector in Polish cities. *Journal of Open Innovation: Technology, Market, and Complexity*, 8, 17.

Wolniak, R., and Stecuła, K. (2024). Evaluation of quality of innovative e-learning in higher education: An insight from Poland. *Applied System Innovation*, 7(6), 109. https://doi.org/10.3390/asi7060109

Wolniak, R., Stecuła, K., and Aydın, B. (2024). Digital transformation of grocery in-store shopping-scanners, artificial intelligence, augmented reality and beyond: A review. *Foods*, 13(18), 2948.

Wolniak, R., Gajdzik, B., Grebski, M., Danel, R., and Grebski, W.W. (2024). Business models used in smart cities—theoretical approach with examples of smart cities. *Smart Cities*, 7, 1626–1669.

Wolniak, R., and Stecuła, K. (2024). Artificial intelligence in smart cities—applications, barriers and future directions: A review. *Smart Cities*, 7, 1346–1389.

Xu, H. (2022). Intelligent automobile auxiliary propagation system based on speech recognition and AI driven feature extraction techniques. *International Journal of Speech Technology*, 25, 893–905.

Yu, Y., Xu, J., Zhang, J.Z., Kamal, M.M., and Cao, Y. (2024). Unleashing the power of AI in manufacturing: Enhancing resilience and performance through cognitive insights, process automation, and cognitive engagement. *International Journal of Production Economics*, 270, 109175.

Zhou, Y. (2022). The application trend of digital finance and technological innovation in the development of green economy. *Journal of Environmental and Public Health*, 2022, 1064558.

CHAPTER 4

AI in Smart Environment

4.1 AI FOR ENVIRONMENTAL MONITORING AND POLLUTION CONTROL

With the progress of civilization and the development of technology, especially industry, the impact of human activity on the environment is becoming more and more noticeable. The technologies introduced increasingly affect the hydrosphere, air quality, soil and flora and fauna, as well as humans and their health. On the other hand, the development of technology can also be used to counteract the negative effects of human activity. A great opportunity is the use of Artificial Intelligence (AI) in the field of controlling and eliminating pollution that enters our environment. AI is poised to play an increasingly important role in environmental monitoring and pollution control, offering modern tools for analysis, forecasting, and decision-making. In the context of Smart Cities, AI-based solutions enable more effective tracking of air quality, noise levels, and other key environmental parameters.

The first example can be the application for air quality modeling. It was used in practice in seven Smart Cities in India (Kapoor et al., 2024). This solution was a part of research which proposes a data-driven approach combining artificial neural networks with particle swarm optimization, a nature-inspired optimization technique, to predict air quality. The study and its results show the importance of accurate air quality modeling for assessing the impact of pollution on living and non-living entities. The authors emphasize how big the role of AI models is when it comes to providing early warnings for high pollutant concentrations. AirQo is another solution that fills in the gaps in air quality data (Bainomugisha et al., 2023). This system features a distributed sensor network in addition to autonomous sensors for deployment in resource-constrained areas. Additionally, there

DOI: 10.1201/9781003587613-5

is a platform for data access and analytics, as well as a network management application that is required for tracking installations and maintenance. In another article, the authors use regression to forecast the quality of the air (Al-Eidi et al., 2023). Three regression models for predicting air quality in Smart Cities are compared in this study: random forest, linear regression, and decision tree. It concludes that the best accuracy and lowest error rates are provided by decision tree regression. The integration of cloud computing significantly improves processing speed, enabling real-time air quality forecasting and faster responses to pollution events. Thanks to distributed computing resources, real-time air quality forecasting becomes possible, enabling timely decision-making and proactive measures to address air pollution episodes effectively. Another study presents the use of AI to predict air quality in six cities in India, one of the most polluted countries in the world according to the IQAir report (Suri et al., 2023). The researchers developed predictive models based on data from the Kaggle platform, covering 6,139 datasets from 2016 to 2020, containing concentrations of ten air pollutants (including $PM_{2.5}$, PM_{10}, NO_2, CO, SO_2). The study compared two machine learning models: an artificial neural network (ANN) and Gaussian process regression (GPR). Both models achieved high predictive performance, with GPR achieving better results ($R = 0.9843$) than ANN ($R = 0.9611$), also in terms of RMSE and MAE errors. The models built enable real-time air quality forecasting, which allows residents to be warned about the threat of smog and support planning of pollution control measures. Another approach involves a predictive system designed to forecast air pollution levels in urban areas (Petry et al., 2021). This approach supports city planners in making timely and effective decisions. It uses sequential encoder-decoder architecture with recurrent neural networks. Its task is to predict concentrations of key pollutants—PM, NO_2, and O_3—for up to 48 hours in advance. The model was trained with historical air quality, traffic, and meteorological data, and the predictions were strongly correlated with real in situ observations. The results confirm that such a method using AI can provide accurate real-time air quality predictions and is therefore a valuable contribution to urban environmental planning as well as management of public health.

There is another technology called LoRa (Ashokkumar et al., 2022). It can be utilized in most Smart City nodes to provide low-power, wide-area communication for smart infrastructure. It offers varied services. For example, street lighting solutions can be controlled centrally by LoRa-based networks, and AI can help automate it by adjusting the schedule of lighting based on time, weather, or traffic. It can also help in providing air quality. In order to ensure air quality and monitor pollution levels, Smart Cities install air quality monitoring systems in strategic areas. These systems, typically backed by AI, continuously sample air quality parameters and assess environmental health to facilitate real-time action and long-term planning. In other studies, researchers also focused on the use of AI and machine learning to monitor air pollution in cities and assess its impact on public health (Krupnova et al., 2022). They used AI/ML models to predict pollutant concentrations and estimate exposure levels of different communities. This approach is important in the context of epidemiological studies and environmental risk analysis. The importance of integrating large environmental datasets and their effective management was emphasized. An important part of huge datasets is their visualization. It is also worth paying attention to the problem of environmental inequality, indicating that access to advanced AI/ML technologies within the Smart City concept is not evenly distributed geographically, which may deepen existing disparities.

When it comes to the environment, another important issue is monitoring and reducing noise, and in this respect, the possible use of AI is also significant. In the literature, we can read about a real-time noise and exhaust emissions monitoring system for sustainable and intelligent transportation systems (Rauniyar et al., 2023). This one is realized under the project called "Noise and Emissions Monitoring and Radical Mitigation (NEMO)." The solution is based on IoT applications and AI algorithms. It gathers vehicle noise and exhaust emissions data in European cities. Then, it categorizes them into different classes of emitters and helps infrastructure managers to take decisions and actions based on this data. As a result, it is possible to monitor and mitigate noise and emissions for transportation systems, having a comprehensive and innovative system. In turn, another solution proposes a dual-branch residual network for environmental sound classification (Zhang et al., 2023). This integrates feature fusion to enhance recognition accuracy. To provide access to technical data, preprocessing—loop-padding to limit the sound efficiency and time-frequency data amplification—allows the use of model overfitting. Experimental results and the application of the extended sound signal method in the urban dataset UrbanSound8k are known from previous models. In other words, the system better recognizes sounds in the city, such as traffic noise, by combining two types of sound analysis and special processing of recordings.

In the era of dynamic development of Smart Cities, more and more attention is paid not only to air quality or energy efficiency, but also to water. Access to clean, safe drinking water is one of the most important human rights. At the same time, it is a big challenge for urbanized, populated areas, where traditional methods of monitoring water quality often are insufficient. In this regard, AI becomes key tool supporting modern water management systems. Lu et al. (2023) in their research propose an early warning system based on a network of online sensors and machine learning algorithms. It enables a quick response in the event of deterioration of water parameters supplied to urban households. Other researchers have created an integrated IT system supporting the quality of river water in urban spaces (Jiang et al., 2023). They propose an integrated approach to urban water quality surveillance (UrbanWQEWIS) that combines monitoring, early warning system, pollution source identification, and emergency response procedures. The researchers developed and implemented a pilot decision support system (EDSS) in the Maozhou River in China using environmental Internet of Things technologies. The key model components of the system were validated based on monitoring data and field experiments. A particular innovation was the combination of fluorescence-based and Bayesian analysis-based pollution source identification methods, which allowed for cost reduction and increased flexibility of the system. The presented solution is a modern tool for water quality management in the digital era and can be used as a model in other regions. The research of another study shows how AI can be used to predict drinking water quality and counteract water pollution (Barzegar et al., 2023). The study proposes the application of a hierarchical fuzzy model for the prediction of Rome's drinking water quality. The Mamdani fuzzy inference system along with different defuzzification methods is used. The model consists of three intermediate fuzzy models and one final model with three input parameters and 27 fuzzy rules each. Input variables are nine physicochemical parameters of water (pH, hardness, iron, nitrates) that should be in the range of acceptable values for human health. Fuzzy logic has proved to be efficient in handling uncertainty and subjectivity in water quality assessment, offering a flexible and easily modifiable tool in support of complex water systems management. In turn, Anjum et al. used GIS tools and machine learning algorithms to assess groundwater quality in the context of Smart Cities (Anjum et al., 2023). All these solutions have one goal: to provide city dwellers with access to healthy water and prevent environmental disasters resulting from its contamination. On the other hand, a slightly different perspective is provided by the other study, who point out the growing need for precise rainfall intensity estimation (Alkhatib et al., 2023). This is claimed especially bearing in mind increasingly frequent and violent weather events as a result from climate change and urbanization. This study presents an innovative approach, which is using audio recordings to estimate rainfall intensity with the application of convolutional neural networks. This solution can become part of Smart City systems and—thanks to the involvement of citizens (citizen science)—also expand the scope of data available for real-time analysis.

4.2 OPTIMIZING ENERGY CONSUMPTION WITH AI

Studies on the application of AI to optimize energy consumption may also be found in both literature and practice. Recent developments in digital technology are essential for reducing energy use, utilizing advanced analytics, and enhancing system efficiency. Innovation, new business strategies, and the creation of energy-efficient goods and services are all facilitated by digitalization (Subramanian, 2023). It encourages the development of low-carbon energy systems that result in notable energy savings as well as Smart Cities, buildings, and houses. These energy-related solutions can affect the city as a whole as well as individual citizens (Stecuła et al., 2023). In order to manage and conserve energy, a variety of techniques are employed, including sensors. For instance, an IoT-based smart metering system is suggested (Li et al., 2024). The proposed system aims to predict energy consumption in Smart Cities based on a set of energy efficiency data and recurrent neural networks. AI and machine learning can be used in non-intrusive load monitoring (NILM) approaches to efficiently monitor and enhance energy usage. Without requiring costly hardware, supervised learning methods such as modified K-Nearest Neighbors and Bagging Regressor may forecast and categorize appliance usage by evaluating energy consumption data from smart meters (Kommey et al., 2024). This method supports smart home decision-making, optimizes energy use, lowers bills, and identifies malfunctioning appliances. All things considered, AI-based NILM promotes the creation of intelligent, sustainable cities and helps with effective energy management.

The importance of smart grids is increasingly emphasized in scientific literature, which plays a key role in energy management in cities of the future. A smart grid, or smart energy network, is a modern infrastructure that combines digital technology with the power supply and distribution system of electricity. Thanks to the use of advanced AI e methods, it is possible not only to manage energy more efficiently and reliably, but also to increase its sustainable use (S. S. Ali and Choi, 2020). The research of Alsolami et al. focuses on the use of deep reinforcement learning to optimize smart grid energy trading and demand response in a decentralized peer-to-peer energy exchange system (Alsolami et al., 2023). The proposed approach, based on decentralized training and performance-based learning, enables efficient management of household energy costs by modeling this complex process as a partially observable Markov decision process. In other studies, the authors use machine learning techniques to analyze abnormal energy consumption patterns in power grids, which is crucial in the context of building green and sustainable cities of the future (Ali et al., 2024). It is also possible to use LSTM (long short-term memory) neural networks and recurrent neural networks (RNN) to improve the adaptive capabilities of modern smart grids, enabling them to better respond to changing conditions of energy consumption and supply (Sulaiman et al., 2023). All this research shows that AI has huge potential to support the energy transformation of cities and create more intelligent, resilient, and efficient energy systems of the future.

It is also worth mentioning the problem of energy theft. In such a case, AI can help detect undesirable behavior. For example, a solution called an Energy Theft Prevention System can be applied (Quasim et al., 2023). This is a multi-objective diagnosing structure which detects energy theft. The solution uses different machine learning techniques: gated recurrent unit, gray wolf optimization, deep recurrent convolutional neural network, long short-term memory. The system identifies irregular energy usage patterns and also enhances the reliability and security of smart energy grids.

When it comes to the area of energy management in residential buildings, systems that use AI play an increasingly important role, which helps in the optimization of energy consumption, cost reduction, and integration with renewable energy sources. Some study presents a model of a home energy management system (HEMS) which is based on two-way communication with the energy supplier. This one optimizes the energy consumption of home appliances. It takes into account information on energy prices and categorizes loads into non-shiftable, shiftable and controllable (Lu et al., 2019). Duman et al. propose an HEMS that takes into account the management of the battery, electric vehicles, and air conditioning, based on data on solar radiation, energy prices, and the presence of residents (Duman et al., 2021). In turn, other research introduces a system based on deep reinforcement learning. Such a system regulates the temperature in rooms and utility water based on energy production from PV, aiming to reduce consumption and increase the comfort of residents (Lissa et al., 2021). Summing up, it can be said that intelligent energy management systems (supported by AI) play a key role in increasing energy efficiency, reducing costs, and integrating renewable energy sources in modern households.

In the context of intelligent energy management, more and more solutions based on AI concern individual users in their homes, especially in the field of HVAC (heating, ventilation, and air conditioning) systems, which AI helps to optimize for comfort and energy efficiency. AI-based solutions allow heating and cooling systems to adapt to weather conditions, the presence of household members, and their preferences, which translates into higher comfort and lower energy consumption (Morse et al., 2021). An example is smart thermostats, which, thanks to machine learning and predictive algorithms, can dynamically manage room temperature (Marantos et al., 2019; Duman et al., 2021). These technologies often work with energy management systems (HEMS), photovoltaics, and electricity price data, and optimize consumption and costs (Huang et al., 2020; Nezhad et al., 2021). Some innovative approaches also include heating control using neural networks and fuzzy logic (Sadeghian Broujeny et al., 2021), as well as adaptive passive systems. These make it possible to automatically switch between cooling and heating modes depending on conditions of environment (An et al., 2022). These solutions are a significant step toward more sustainable and energy-efficient smart homes.

There are also intelligent lighting systems in smart homes. Smart lighting systems in smart homes use sensors, IoT, and AI to automatically adjust the light level depending on the presence of household members, time of day, or amount of natural light, which translates into energy savings and increased user comfort. One example is the system that Chiesa et al. describe, which uses fuzzy logic for a quick and effective response and combines lighting with a dynamic shading system based on Arduino, Raspberry Pi, and a mobile application (Chiesa et al., 2020). A lighting control system based on inexpensive wireless modules is shown in another study. It enables automatic shut-off of lights and other electrical equipment in smart home architecture (Andramuño et al., 2021). The personalization of lighting using information from lifelogs is another interesting strategy. Such solution collects and combines information about user activities, emotions, and the environment to produce customized lighting sceneries (Cho et al., 2020). A study by Ayan and Turkay showed that smart LED bulbs, especially when using dimming and remote-control functions, can really reduce energy consumption compared to traditional light sources (Ayan and Turkay, 2020). To summarize, smart lighting systems using AI and IoT on the one hand increase the comfort and safety of residents, and on the other hand play an important role in improving the energy efficiency of modern homes.

Due to the fact that AI has been developing at a very fast pace in recent years, it is possible to identify certain future directions in which AI can help optimize energy consumption. Table 4.1. presents the future directions and their description, highlighting the benefits.

4.3 AI IN WASTE MANAGEMENT AND CLIMATE RESILIENCE

The use of AI in climate waste management is becoming increasingly important in the context of sustainable development and environmental protection. AI enables the automation and optimization of waste segregation processes, which translates into higher recycling efficiency and a reduction in the amount of waste going to landfills. AI-based systems can analyze the composition of waste, identify materials suitable for recovery,

TABLE 4.1 Future Directions of Optimizing Energy Consumption with AI

Future Direction	Description	Benefits
Predictive energy management	Forecasting energy consumption based on weather data, energy prices, and usage patterns	Lower energy costs, efficient resource management
Integration with photovoltaics and energy storage	Matching energy use with current production and storage capacity	Increased self-consumption, reduced grid load
Personalized thermal comfort settings	Automatically adjusting temperature based on user preferences and presence	Higher comfort, reduced HVAC energy use
Optimal charging of electric vehicles	Scheduling EV charging based on tariffs and renewable energy availability	Lower operating costs, better energy utilization
Anomaly and device fault detection	Monitoring energy used to detect inefficient devices or system failures	Faster diagnostics, prevention of energy losses
Smart lighting control	Automatically turning on/off or dimming lights based on occupancy and time of day	Energy savings, improved user convenience
Real-time energy usage profile management	Dynamically adjusting consumption based on pricing, demand, and resource availability	Cost and emission reduction, peak demand relief

Source: Author's own work.

and predict their quantity, which helps in better planning of logistics and operations related to waste management.

AI systems can recognize plastics, glass, paper, and metals in trash images on sorting belts and route them to the proper recycling fractions by using technologies like computer vision and machine learning (Olawade et al., 2024). These systems boost sorting efficiency, eliminate human intervention in the process, increase recycling efficiency, and decrease landfill waste. Large data sets may be analyzed in real time by machine learning, which makes it possible to predict garbage production, optimize collection routes, expedite the sorting and recycling procedures, and pinpoint pollution sources (Rautela et al., 2025). This allows for better planning of activities and decision-making in the field of environmental protection. AI also supports the development of waste-to-energy technologies, by increasing the efficiency of energy recovery and reducing greenhouse gas emissions. AI has the potential to be a useful instrument in accomplishing the Sustainable Development Goals (SDGs) when paired with legislative frameworks and environmental regulations. However, more research and use of these methods are still needed to address issues including algorithm bias, data security, and the requirement for interdisciplinary collaboration. Another study focuses on classifying waste materials to be used in automated sorting systems that eliminate manual waste segregation (Tehrani and Karbasi, 2017). Such research can use ANNs to recognize various waste fractions. In detail, hyperspectral imaging and multi-layer ANNs were applied to recognize different types of plastics in e-waste. Chu et al. used convolutional neural networks for feature extraction and multilayer perceptron to classify waste into recyclable and non-recyclable (Chu et al., 2018). Ysabel et al. used data mining techniques (including cluster analysis and decision tree classifier) to correlate social and behavioral factors with waste generation (Ysabel et al., 2008). Sudha et al. conducted research on how to automate the

TABLE 4.2 Examples of AI Usage in Climate Resilience

Area	Example
Agriculture	AI analyzes weather and soil data to predict droughts and optimize planting dates
Water management	AI predicts flood risk based on rainfall and topographic data
Cities (urban resilience)	AI systems monitor air quality and control traffic during heat waves
Healthcare	AI detects increased risk of infectious diseases after extreme weather events
Energy	AI optimizes power grid performance during heat-related overloads
Natural disasters	AI analyzes satellite imagery and meteorological data to predict earthquakes and hurricanes
Noise and pollution monitoring	AI analyzes sensor data to detect and predict environmental violations
Biodiversity protection	AI analyzes data from cameras and drones to protect ecosystems threatened by climate change
Natural hazard prediction	AI processes satellite and climate data to predict forest fires and landslides in urban areas
Spatial planning	AI supports the location of new investments taking into account climate risks (e.g., floods, landslides, heat)

Source: Author's own work.

garbage sorting process with the use of deep Convolutional Neural Network (CNNs) (Sudha et al., 2016). The same technology was applied to separate different types of paper and cardboard in some other research (Vrancken et al., 2019).

AI supports modeling and forecasting the effects of climate change, which allows for more effective preparation for extreme weather events and their minimization. AI analyzes huge sets of climate data, monitors trend and assesses the risk of floods, droughts or fires, thus supporting the planning of climate-resilient infrastructure and adaptation actions at the local and regional level. Examples are presented in Table 4.2.

By integrating AI into climate tools, it is possible not only to increase operational efficiency, but also to make more informed decisions that promote environmental protection and increase the resilience of communities to the negative effects of climate change.

BIBLIOGRAPHY

Al-Eidi, S., Amsaad, F., Darwish, O., Tashtoush, Y., Alqahtani, A., and Niveshitha, N. (2023). Comparative analysis study for air quality prediction in smart cities using regression techniques. *IEEE Access*, 11, 115140–115149. https://doi.org/10.1109/ACCESS.2023.3323447

Ali, A., Khan, L., Javaid, N., Aslam, M., Aldegheishem, A., and Alrajeh, N. (2024). Exploiting machine learning to tackle peculiar consumption of electricity in power grids: A step towards building green smart cities. *IET Generation, Transmission and Distribution*, 18(3), 413–445. https://doi.org/10.1049/gtd2.13056

Ali, S.S. and Choi, B.J. (2020). State-of-the-art artificial intelligence techniques for distributed smart grids: A review. *Electronics*, 9(6), 1030.

Alkhatib, M.I.I., Talei, A., Chang, T.K., Pauwels, V.R.N., and Chow, M.F. (2023). An urban acoustic rainfall estimation technique using a CNN inversion approach for potential smart city applications. *Smart Cities*, 6(6), 3112–3137.

Alsolami, M., Alferidi, A., Lami, B., and Ben Slama, S. (2023). Peer-to-peer trading in smart grid with demand response and grid outage using deep reinforcement learning. *Ain Shams Engineering Journal*, 14(12). https://doi.org/10.1016/j.asej.2023.102466

An, Y., Fu, Y., Dai, J.-G., Yin, X., and Lei, D. (2022). Switchable radiative cooling technologies for smart thermal management. *Cell Reports Physical Science*, 3(10), 1–11.

Andramuño, J., Mendoza, E., Núñez, J., and Liger, E. (2021). Intelligent distributed module for local control of lighting and electrical outlets in a home. *Journal of Physics: Conference Series*, 1730(1), 12001.

Anjum, R., Ali, S.A., and Siddiqui, M.A. (2023). Assessing the impact of land cover on groundwater quality in a smart city using GIS and machine learning algorithms. *Water, Air, and Soil Pollution*, 234(3). https://doi.org/10.1007/s11270-023-06198-8

Ashokkumar, S.R., Harihar, R., Subhashini, R., and Naveen Prasaath, S. (2022). A literature survey on artificial intelligence-based smart city automation using LoRa and IOT for street lights and air quality check. *2022 1st International Conference on Computer, Power and Communications, ICCPC 2022 – Proceedings*, pp. 149–152. https://doi.org/10.1109/ICCPC55978.2022.10072071

Ayan, O., and Turkay, B. (2020). IoT-based energy efficiency in smart homes by smart lighting solutions. *2020 21st International Symposium on Electrical Apparatus & Technologies (SIELA)*, pp. 1–5.

Bainomugisha, E., Ssematimba, J., and Okure, D. (2023). Design considerations for a distributed low-cost air quality sensing system for urban environments in low-resource settings. *Atmosphere*, 14(2). https://doi.org/10.3390/atmos14020354

Barzegar, Y., Gorelova, I., Bellini, F., and D'Ascenzo, F. (2023). Drinking water quality assessment using a fuzzy inference system method: A case study of Rome (Italy). *International Journal of Environmental Research and Public Health*, 20(15). https://doi.org/10.3390/ijerph20156522

Chiesa, G., Di Vita, D., Ghadirzadeh, A., Herrera, A. H. M., and Rodriguez, J.C.L. (2020). A fuzzy-logic IoT lighting and shading control system for smart buildings. *Automation in Construction*, 120, 103397.

Cho, Y., Seo, J., Lee, H., Choi, S., Choi, A., Sung, M., and Hur, Y. (2020). Platform design for lifelog-based smart lighting control. *Building and Environment*, 185, 107267.

Chu, Y., Huang, C., Xie, X., Tan, B., Kamal, S., and Xiong, X. (2018). Multilayer hybrid deep learning method for waste classification and recycling. *Computational Intelligence and Neuroscience*. 2018, 9.

Duman, A. C., Erden, H. S., Gönül, Ö., and Güler, Ö. (2021). A home energy management system with an integrated smart thermostat for demand response in smart grids. *Sustainable Cities and Society*, *65*, 102639.

Huang, K., Hallinan, K.P., Lou, R., Alanezi, A., Alshatshati, S., and Sun, Q. (2020). Self-learning algorithm to predict indoor temperature and cooling demand from smart WiFi thermostat in a residential building. *Sustainability*, 12(17), 7110.

Jiang, J., Men, Y., Pang, T., Tang, S., Hou, Z., Luo, M., Sun, X., Wu, J., Yadav, S., Xiong, Y., Liu, C., and Zheng, Y. (2023). An integrated supervision framework to safeguard the urban river water quality supported by ICT and models. *Journal of Environmental Management*, 331. https://doi.org/10.1016/j.jenvman.2023.117245

Kapoor, N.R., Kumar, A., Kumar, A., Kumar, A., and Arora, H.C. (2024). Air quality modeling for smart cities of India by nature inspired AI—A sustainable approach. In: A. Biswas, A.P. Tonda, R. Patgiri, & K.K. Mishra (Eds.), *Advances in Computers*, vol. 135. Cambridge, MA: Elsevier, pp. 129–154.

Kommey, B., Tamakloe, E., Kponyo, J.J., Tchao, E.T., Agbemenu, A.S., and Nunoo-Mensah, H. (2024). An artificial intelligence-based non-intrusive load monitoring of energy consumption in an electrical energy system using a modified K-Nearest Neighbour algorithm. *IET Smart Cities*. https://doi.org/10.1049/smc2.12075

Krupnova, T.G., Rakova, O.V, Bondarenko, K.A., and Tretyakova, V.D. (2022). Environmental justice and the use of artificial intelligence in urban air pollution monitoring. *Big Data and Cognitive Computing*, 6(3). https://doi.org/10.3390/bdcc6030075

Li, X., Zhao, H., Feng, Y., Li, J., Zhao, Y., and Wang, X. (2024). Research on key technologies of high energy efficiency and low power consumption of new data acquisition equipment of power Internet of Things based on artificial intelligence. *International Journal of Thermofluids*, 21, 100575.

Lissa, P., Deane, C., Schukat, M., Seri, F., Keane, M., and Barrett, E. (2021). Deep reinforcement learning for home energy management system control. *Energy and AI*, 3, 100043.

Lu, H., Ding, A., Zheng, Y., Jiang, J., Zhang, J., Zhang, Z., Xu, P., Zhao, X., Quan, F., Gao, C., Jiang, S., Xiong, R., Men, Y., and Shi, L. (2023). Securing drinking water supply in smart cities: An early warning system based on online sensor network and machine learning. *Aqua Water Infrastructure, Ecosystems and Society*, 72(5), 721–738. https://doi.org/10.2166/aqua.2023.007

Lu, R., Hong, S.H., and Yu, M. (2019). Demand response for home energy management using reinforcement learning and artificial neural network. *IEEE Transactions on Smart Grid*, 10(6), 6629–6639.

Marantos, C., Lamprakos, C., Siozios, K., and Soudris, D. (2019). Towards Plug&Play smart thermostats for building's heating/cooling control. In: K. Siozios, D. Anagnostos, D. Soudris, & E. Kosmatopoulos (Eds.), *IoT for Smart Grids: Design Challenges and Paradigms*. Cham: Springer, pp. 183–207.

Morse, H.F., AlGhamdi, M.S., Almusaied, H.M., Alsubaie, S.A., Almarri, D.M., Alqahtani, M.A., Alqahtani, R.A., Tabeidi, R.A., and Balharith, T. (2021). The smart heater based on internet of things. *The International Conference on Artificial Intelligence and Computer Vision*, pp. 145–153.

Nezhad, A.E., Rahimnejad, A., and Gadsden, S.A. (2021). Home energy management system for smart buildings with inverter-based air conditioning system. *International Journal of Electrical Power & Energy Systems*, 133, 107230.

Olawade, D.B., Fapohunda, O., Wada, O.Z., Usman, S.O., Ige, A.O., Ajisafe, O., and Oladapo, B.I. (2024). Smart waste management: A paradigm shift enabled by artificial intelligence. *Waste Management Bulletin*, 2(2), 244–263. https://doi.org/https://doi.org/10.1016/j.wmb.2024.05.001

Petry, L., Meiers, T., Reuschenberg, D., Mirzavand Borujeni, S., Arndt, J., Odenthal, L., Erbertseder, T., Taubenböck, H., Müller, I., Kalusche, E., Weber, B., Käflein, J., Mayer, C., Meinel, G., Gengenbach, C., and Herold, H. (2021). Design and results of an AI-based forecasting of air pollutants for smart cities. *ISPRS Annals of the Photogrammetry, Remote Sensing and Spatial Information Sciences*, 8(4/W1-2021), 89–96. https://doi.org/10.5194/isprs-annals-VIII-4-W1-2021-89-2021

Quasim, M.T., Nisa, K.ul, Khan, M.Z., Husain, M.S., Alam, S., Shuaib, M., Meraj, M., and Abdullah, M. (2023). An internet of things enabled machine learning model for Energy Theft Prevention System (ETPS) in Smart Cities. *Journal of Cloud Computing*, 12(1), 158.

Rauniyar, A., Berge, T., Kuijpers, A., Litzinger, P., Peeters, B., Gils, E. Van, Kirchhoff, N., and Hakegard, J.E. (2023). NEMO: Real-time noise and exhaust emissions monitoring for sustainable and intelligent transportation systems. *IEEE Sensors Journal*, 23(20), 25497–25517. https://doi.org/10.1109/JSEN.2023.3312861

Rautela, K.S., Goyal, M.K., and Surampalli, R.Y. (2025). AI and machine learning for optimizing waste management and reducing air pollution. *Journal of Hazardous, Toxic, and Radioactive Waste*, 29(3), 4025014.

Sadeghian Broujeny, R., Madani, K., Chebira, A., Amarger, V., and Hurtard, L. (2021). A heating controller designing based on living space heating dynamic's model approach in a smart building. *Energies*, 14(4). https://doi.org/10.3390/en14040998

Stecuła, K., Wolniak, R., and Grebski, W.W. (2023). AI-driven urban energy solutions—from individuals to society: A review. *Energies*, 16(24), 7988.

Subramanian, M. (2023). Leveraging digitalization for improving energy efficiency. In: *The Handbook of Energy Policy*. https://doi.org/10.1007/978-981-19-6778-8_22

Sudha, S., Vidhyalakshmi, M., Pavithra, K., Swaathi, V., and Sangeetha, K. (2016). An automatic classification method for environment. In: *IEEE International Conference on Technological Innovations in ICT for Agriculture and Rural Development (TIAR 2016)*, pp. 65–70.

Sulaiman, A., Nagu, B., Kaur, G., Karuppaiah, P., Alshahrani, H., Reshan, M. S. Al, AlYami, S., and Shaikh, A. (2023). Artificial intelligence-based secured power grid protocol for smart city. *Sensors*, 23(19), 8016.

Suri, R.S., Jain, A.K., Kapoor, N.R., Kumar, A., Arora, H.C., Kumar, K., and Jahangir, H. (2023). Air quality prediction: A study using neural network based approach. *Journal of Soft Computing in Civil Engineering*, 7(1), 93–113. https://doi.org/10.22115/SCCE.2022.352017.1488

Tehrani, A., Karbasi, H. (2017). A novel integration of hyper-spectral imaging and neural networks to process waste electrical and electronic plastics. *IEEE Conference on Technologies for Sustainability (SusTech)*, 5.

Vrancken, C., Longhurst, P., Wagland, S. (2019). Deep learning in material recovery: Development of method to create training database. *Expert Systems with Applications*, 125, 268–280. https://doi.org/10.1016/j.eswa.2019.01.077

Ysabel, M., Ojeda, S., and Hidalgo, H. (2008). Identification of behavior patterns in household solid waste generation in Mexicali's city: Study case. *Resources, Conservation and Recycling*, resconrec.2008.07.011.

Zhang, D., Zhong, Z., Xia, Y., Wang, Z., and Xiong, W. (2023). An automatic classification system for environmental sound in smart cities. *Sensors*, 23(15). https://doi.org/10.3390/s23156823

CHAPTER 5

AI in Smart Mobility

5.1 INTELLIGENT TRANSPORTATION SYSTEMS

Smart mobility is the combined application of new technologies to enhance transport systems. It includes smart traffic, public transport optimization, and green modes of transport like electric vehicles and shared transport like car-sharing or bicycle-sharing systems. Maximizing city mobility, reducing congestion, and minimizing the environmental impact of transport are the main goals (McKinsey Global Institute, 2018; Belaïd et al., 2023).

Incorporating Artificial Intelligence (AI) into intelligent mobility solutions is needed to create efficient, sustainable, and well-integrated transport systems that are popular in Smart Cities. AI manages traffic effectively and removes congestion by analyzing vast amounts of data in real-time (Leali and Pasquale, 2022; Alanazi, 2023; Seng et al., 2023). Traffic management systems based on AI can predict traffic trends, detect congestion spots, and change signal phases dynamically, thereby de-congesting the system using sophisticated algorithms (Moumen et al., 2023). In addition to optimizing transport systems, this approach reduces fuel consumption and also cuts down greenhouse gases (Abbas et al., 2023). The heterogeneity and complexity of the urban transport system should be brought to the fore as well. There are some researches (Musa et al., 2023) that provide the synergy of Internet of Things (IoT) and Intelligent Transportation System (ITS) solutions as a remedy for Smart City traffic congestion. Driverless as well as human-operated cars are surrounded by real-time traffic streams as well as AI sensor and ITS hardware information in such scenarios. Data collected is analyzed by cloud computing and machine learning. These steps guarantee effective traffic management, policy-making, and documentation for future use. AI can be utilized in the development of predictive

DOI: 10.1201/9781003587613-6

maintenance models for public transport infrastructure (Kamran et al., 2023). According to bus, train, and other vehicle sensor surveys, AI software can forecast and forecast the maintenance required, reducing downtime and public transport reliability (Maldonado Silveira Alonso Munhoz et al., 2020). There are many maintenance approaches, but AI is much promising as per what has been obtained from literature (Wieczorek et al., 2024).

Following is the AI-dependent and shared mobility-related content. Shared mobility is a new concept of urban mobility in which users share usage of cars, bicycles, scooters, and other vehicles one after the other or as combinations (Machado et al., 2018). Technology is employed to provide agile, affordable, and sustainable modes of transport compared to traditional car ownership. Internet-based services and applications enable customers to book, use, and pay for use at low and affordable prices. Benefits include reduced traffic on the road, reduced transport expenses for consumers, reduced greenhouse gas emissions, and increased efficient use of available transport infrastructure (Turoń and Tóth, 2023). Shared mobility also improves public transport networks by enabling last-mile connections, thereby guaranteeing greater overall urban mobility and more people-oriented cities (Santos, 2018). AI solutions directly developed for machine learning uses are used for the optimization of shared mobility system management. For example, it is demonstrated by the author's model (Kubik, 2023) that it is possible to classify vehicle-based trips from shared mobility systems correctly. Its applicability in daily life is not just functional in the meaning of making management processes simpler but also due to its simplicity in being updated to cater to changes in the future. Alternatively, a concept of smart biking is increasingly in demand. Smart cycling and its application in Smart Cities are the subject of current publications (Wolniak, 2023c; Wolniak and Grebski, 2023). Worthy of note is that intelligent mobility platforms, built on AI, make modes of transport easily integrable, i.e., public transport, ride-sharing, and bike-sharing are on the cards. They give people minute-by-minute real-time information and optimize travel planning, involving diverse modes of transport (Al Mamlook et al., 2023; Turoń, 2023; Turoń and Tóth, 2023). AI technology takes into consideration individual preferences, previous data, and present conditions in an attempt to propose customized and effective ways of transport, enhancing city mobility according to client needs and adaptability (Rosa et al., 2022).

AI deployment in smart mobility is not merely a matter of reacting to the youth population's needs and demands but also in relation to urban sustainability and safety of the elderly population. The elderly will typically have another mobility problem, such as lower bodily ability and higher susceptibility to accidents. AI technologies are increasingly being used to help tackle these issues proportionally. At least the most intriguing among intelligent mobility applications of AI is to optimize transport accessibility and security of older people's transportation. AI innovations also bring about accessible transport services, particularly tailored to older people (Vechione et al., 2018). For example, intelligent routing technologies can suggest available routes that do not have physical barriers and are not steep slopes to allow older adults to move freely and independently. AI is also capable of making transport safer through monitoring road conditions and alerting drivers and pedestrians about potential dangers (Oksman, 2018). Bokolo (2023) argues that safety and accessibility within public transport systems should incorporate cutting-edge technologies like AI-based services. These emerging technologies aim to offer personalized advice enhancing the mobility experience as well as the safety elements, mainly to vulnerable users such as older citizens above 65. The research emphasizes the promise of emerging technologies such as AI-driven machine learning for transforming the accessibility and safety of transport services offered to the elderly.

Another segment of Smart Cities where AI is observed and constructed is electric cars. Electric cars are a core business in constructing Smart Cities, offering green transport solutions along with green concerns. Electric vehicles (EVs) in Smart Cities are constructed into end-to-end mobility plans that concentrate on decreasing carbon emissions and enhancing air quality. AI technology makes it possible to install EV charging facilities with optimal charging station locations and optimal power distribution. AI can be a fundamental element in electric vehicle charging facilities to enhance its efficiency, reliability, and user satisfaction (Sun et al., 2020). Machine learning models can also learn parameters like the grid capacity, prices of energy, and behavioral patterns of users to calculate optimal charge schedules for electric vehicles (Ahmed et al., 2021). Charging is therefore performed during off-peak hours or when there is a surplus of renewable energy over demand, thereby minimizing cost and emissions (Raza et al., 2022). Apart from this, AI facilitates demand response programs in which charging time of EV customers is synchronized to overlap with grid stability. During peak load, this avoids likely outages (Bajaj and Siddharth, 2023). Other researchers (ElHusseini et al., 2020) suggested using blockchain technology and AI technology. Combining both the technologies has the potential to enhance charging in terms of security, efficiency, and decentralization. Others of them are found in textbooks, for example (Ahmed et al., 2021; Paret et al., 2023). Their benefit might be increasing security and optimizing charging (Swarnkar et al., 2022).

As observed in other studies (Li et al., 2021), EV AI can train a good prediction model in an effort to schedule chargers. Such prediction models enable more data-driven planning and management of resources. In solution phase, the AI plays most critical role of competitive resource optimization in a way to utilize within charger planning. This includes strategic choice on charging point position and assigning them optimally based on predicted demand. AI calculation is utilized to arrive at the correct choice on EV charging gear size so that correct quantities of charging points get installed at different places of a city. Such a forward-looking strategy not only maximizes supply of charging stations but also helps achieve smooth integration of electric vehicles in the transport network of urban regions. Besides this, AI can be used to forecast EV charging and discharging patterns (Chen and Folly, 2022). Another study (Mosayebi et al., 2022) proposed an intelligent extreme fast portable charger for electric vehicles based on multiple input sources such as the grid and other alternative energy systems like energy storage. The method is intended to optimize charging efficiency through learning to available power sources and EV battery health using a model-free sliding mode controller. The process is designed to lengthen the battery and system efficiency lifespan, as verified from real-time tests. AI enables the creation of three-stage vehicle-to-grid (V2G) technology (Dharavat et al., 2023). During the first stage, EV charging is regulated through uncontrolled and controlled approaches. The second stage, with greater EV penetration, requires intelligent charging/discharge control and aggregator coordination. The third phase foresees enormous EV support in delivering ancillary services to power networks (Cao et al., 2023).

In electric vehicles, there needs to be a comparison with combustion vehicles. While the trend appears to be investing in low-emission or zero-emission vehicles, the combustion vehicle problem still exists and must be tackled. The initial avenue of solving this problem is AI, specifically emissions-wise. Its uses are not just limited to vehicle emissions but also company and house space emissions. AI technologies are at the vanguard of maximizing the use of energy, minimizing inefficiency, and limiting environmental impact in industries to aid global initiatives toward sustainability (Zhao et al., 2021). AI significantly contributes to the reduction of emissions from vehicles by maximizing the efficiency of the traditional internal combustion engine. AI software code stored in engine control units monitors engine temperature, load, and air-fuel mixture parameters in real time.

This real-time tracking allows effective tuning for maximum combustion efficiency and minimum emissions (Silva and Martins, 2020). Eco-driving coaches based on AI utilize real-time vehicle information to deliver drivers actionable suggestions that will minimize fuel consumption and emissions (Kumar et al., 2023). They monitor the speed of the engine, performance, and fuel usage so that they can choose optimal driver behavior in terms of optimal speeds, points of shifting, and acceleration and deceleration behaviors (Nousias et al., 2019). All this assists in decreasing carbon footprint and, therefore, assist in decreasing emissions as well as facilitate shift from electric to internal combustion modes (Ghaffarpasand et al., 2022). In addition, they also do route planning to avoid congestion and unnecessary driving, fuel-economizing vehicles later on, and lowering emissions (Tselios et al., 2019). AI products play a significant role in optimizing road capacity and avoiding congestion, and hence emissions. Traffic patterns are analyzed through real-time traffic feed from traffic cameras, sensors, and smartphone apps, using which AI-based traffic management systems work (Boru İpek, 2023). These networks can dynamically control traffic lights, offer route options, and apply variable toll charge policies during peak hours to reduce congestion and optimize traffic flow (Mandžuka, 2019). Not only does this enhance the mobility of the city but also enhance environmental friendliness by having low emissions of idling and stop-and-go vehicles.

5.2 AUTONOMOUS VEHICLES AND AI-ENHANCED TRAFFIC MANAGEMENT

But another response under Smart Mobility is the autonomous cars chapter. Autonomous cars are a future promise. But there are some short-term solutions that assist the car driver to an extent due to AI (Duggal et al., 2021). The technology will change urban mobility also by ensuring safety, reducing accidents, and improving transport overall (Sawhney, 2023; Tarawneh et al., 2023).

Self-driving cars are included in Smart City planning and administration, i.e., integration of emerging technology development and city planning for improved efficiency, sustainability, and quality of life. Integration of Autonomous Vehicles (AV) in cities has transport, logistics, security, and environmental sustainability dimensions (Amir and Haque, 2024; Ma et al., 2024).

Public transportation is one of the applications of autonomous cars in Smart Cities. Autonomous buses and shuttles will also be able to provide affordable, efficient, and convenient mobility services, especially in underdeveloped areas with poor conventional public transport systems. The buses can follow predefined routes or dynamically adjust routes based on the needs of passengers, hence reducing route planning and waiting time (Louati et al., 2024). The employment of AVs in public transit can also result in a significant reduction in operating expenses through minimizing the need for human drivers and facilitating 24-hour service without the hindrance of human exhaustion (Louati et al., 2024).

In addition to public transit, autonomous cars are revolutionizing urban logistics and delivery operations. Autonomous delivery rovers and autonomous delivery trucks can potentially enhance the last-mile delivery system better with fewer traffic jams and emissions from traditional delivery trucks. Intelligent route algorithms and real-time information with these automobiles can provide them the ability to drive in the city intelligently with efficient movement of goods and services and high dependability. It is even more efficient in high-density urban cities and areas with population density, and where traffic jam will be a costly burden to pay for delivery time (Dodia et al., 2023; Saini et al., 2023).

Safety advantages of AVs are yet another high-priority sector of their application in Smart Cities. Thanks to cutting-edge sensors, machine learning algorithms, and real-time processing power, AVs are reducing accidents caused by human mistakes to a very great extent. Lane departure warning, collision avoidance, and adaptive cruise control are only some of the safety features that are enhancing road safety and even saving lives. In addition, worldwide use of AV reduces emergency care and medication for disease and injury resulting from road accidents once more, freeing city infrastructure from the pressure (Aurangzeb et al., 2024).

Green ecology is among the fundamental ideas of Smart City projects, and AV are making it a success in many ways. Most significantly, electric AVs have no tailpipe emissions at all, thereby decreasing urban air pollution along with greenhouse gas emissions. Efficiencies also translate to traffic flow when AV are used to manage it, resulting in reduced fuel consumption and emissions from conventional internal combustion engine vehicles. By reducing the amount of vehicles on the road and encouraging shared mobility solutions, AVs reduce the environmental impact of city mobility (Tekouabou, 2021; Zhou et al., 2021).

The incorporation of AV in Smart City environments also makes it easier to introduce smart traffic management systems. They make use of AV data to regulate and manage traffic around the clock, synchronizing traffic light cycles and actively redirecting the flow of traffic in order to avoid congestion and improve flow. Integration of AV and traffic control systems can raise the overall productivity in transport as well as reduce the travel time of commuters (Subramanian et al., 2023; Maniyar et al., 2024).

You can even take it a step further and say application of AVs is even bestowed upon carriage of goods and human beings to urban work like street cleaning, trash collection, and road repair. All these functions can be carried out by autonomous cars with great precision and effectiveness in a way that provides city services in the most appropriate way at reduced human labor and operation expenses. This is in the larger context of making city spaces more responsive and efficient (Patel et al., 2023).

AV rollout in Smart Cities relies on the strong digital infrastructure, including high-speed internet connectivity, smart sensors, and data analytics software. Infrastructure provides for smooth AVs operation such that they are capable of communicating with one another as well as with city management systems. The data collected and examined on AVs is reflective of metropolitan mobility patterns in metropolises that can be applied toward the development of cities in the future as well as policymaking (Zhou et al., 2021; Dodia et al., 2023).

In Table 5.1, there is a description of main usages of AV in Smart Cities. In Table 5.2, there is a description of methods used in AI-enhanced traffic management in Smart City.

AI can even be used to help paragraphs about parking automobiles, as well as leaving and locking cycles and scooters in certain areas of the city. Sensor-parking bays with AI-software instruct drivers to free spaces without difficulty, disposing of traffic jams and lowering the green footprint of futile circling (Piccialli et al., 2021; Lee et al., 2023).

In car parking, AI technology can be used to optimize the entire process, ranging from locating empty parking lots to space optimization and safety. High-level algorithms can scan real-time inputs from sensors, cameras, and others to identify empty parking lots and direct drivers to them in an efficient manner, reducing wasted time in parking and minimizing traffic congestion. In addition, AI could provide dynamic pricing systems, with fluctuating parking fees based on demand and supply, to assist in optimizing maximum best utilization of space as well as collecting more revenue for the city governments. In addition, AI-based surveillance systems can enhance safety for parked vehicles on the road by monitoring parking lots for unsafe activity and alerting authorities in the event of any possible threat, thereby providing an added security against theft and vandalism (Manfletti et al., 2023; Almatar, 2024).

AI can also turn out to be a great assistance in making cities better planned and scooter- and bike-friendly. AI intelligent docking stations can store and retrieve such vehicles more efficiently. Machines will be able to monitor real-time available empty docking capacity, and customers will be able to identify where they can drop their bikes or scooters (Wolniak, 2023b, 2023a). AI may be utilized in predictive maintenance as well to monitor the health of the docking stations and scooters, ensuring they function properly and reducing repair time.

Another sector where AI may prove to be an invaluable assistance is security. AI-based surveillance and monitoring systems can ensure theft and vandalization of scooters and

TABLE 5.1 The Main Area of Usage of Autonomous Vehicles in Smart Cities

Usage	Description
Public transportation	Autonomous shuttles and buses offer economical, dependable, and efficient transport options. They can be dynamically re-routed or fixed routed based on passenger demand, maximizing on route planning and waiting time. With decreased human-driver dependence comes decreased cost operations and the possibility of continuous operation with no human-fatigue limit.
Urban logistics and delivery services	Self-driving delivery vans and drones make the last-mile delivery efficient, cutting congestion and emissions of traditional delivery trucks. Real-time information and intelligent routing algorithms allow the vehicles to navigate the city with least possible congestion, faster and more reliable delivery of goods and services, especially in cities.
Traffic safety	With high-end sensors and machine learning capabilities, self-driving cars can mitigate traffic accidents due to human error by a large margin. Technologies like collision avoidance, lane departure warning, and adaptive cruise control improve road safety, and it is possible to save lives and decrease emergency response services and medical treatment for traffic accidents.
Environmental sustainability	Electric autonomous cars do not have any tailpipe pollution, and greenhouse gases and urban air pollution are minimized. Optimizing the traffic flow with AVs means the fuel use and emissions associated with the traditional vehicles will go down. Their shared mobility services minimize the environmental impact of urban transportation by reducing the overall number of cars on the road.
Intelligent traffic management	Autonomous cars enable intelligent traffic management systems to be designed using data to observe and manage traffic in real time. The system can control traffic lights and dynamically modify traffic flow to improve efficiency and alleviate congestion, making the overall transport efficient and minimizing travel time.
Municipal services	Autonomous transportation is extended to city services like garbage collection, street sweeping, and maintenance of infrastructure. They perform them with the highest degree of precision and efficiency, thus delivering the maximum amount of service without applying much human labor and operational expenditure. Examples of this include more responsive and efficient cities.
Data-driven urban planning	The technology supporting self-driving vehicles allows for the measurement and examination of large amounts of data related to city mobility. These insights are used to guide smart, more effective cities through informed, data-driven policy-making and future-oriented urban planning.
Reduced operational costs	Deployment of driverless cars in the majority of application, from buses to urban services, minimizes the demand for human labor, thus saving huge amounts of money. Low operating costs are realized through continuous operation ability and operational efficiency enhancements such that city budgets could be spent more efficiently.
Enhanced mobility access	Autonomous Vehicles can offer better mobility solutions for people with disabilities, older adults, and others whose mobility is restricted to public means of transport. By offering on-demand and accessible transportation, AVs lead to higher levels of inclusion and independence of all citizens in cities.
Emergency response	Autonomous Vehicles can be placed into emergency response networks, giving prompt and accurate transportation of medical equipment, supplies, and personnel. Their capacity to optimize movement and navigation with security can enhance emergency response time and efficiency, enhancing urban populations' resilience and security.
Parking management	AVs have the potential to cut the parking needs of a city by a huge amount by dropping off their riders and parking somewhere else where land is less constrained. This optimal use of space has the potential to release needed city land for alternative purposes, including parks, walk areas, or further development, that can help make a city more human-friendly.

Source: Author's own work.

TABLE 5.2 The Methods Used for AI-Enhanced Traffic Management in Smart Cities

Method	Description
Adaptive traffic signal control	This method involves using AI algorithms to dynamically adjust the timing of traffic lights based on real-time traffic conditions. By analyzing data from sensors and cameras, the system can optimize traffic flow, reduce congestion, and minimize waiting times at intersections.
Predictive traffic modeling	AI-driven predictive modeling analyzes historical and real-time traffic data to forecast future traffic patterns. This enables city planners and traffic management centers to anticipate congestion and implement preemptive measures such as adjusting traffic signal timings or rerouting traffic.
Incident detection and response	AI systems can monitor traffic in real-time to detect accidents, breakdowns, or other incidents. These systems can quickly notify emergency services and provide alternative route suggestions to drivers, thereby minimizing disruptions and ensuring a rapid response to incidents.
Smart parking management	This method uses AI to identify available parking spaces and guide drivers to them. By analyzing data from sensors in parking lots and along streets, the system can reduce the time spent searching for parking, decrease congestion, and optimize the utilization of parking resources.
Traffic flow optimization	AI analyzes data from various sources, including GPS devices, cameras, and social media, to optimize traffic flow. This can involve adjusting traffic signal timings, opening or closing lanes, and providing real-time traffic updates to drivers to improve overall traffic efficiency and reduce travel times.
Vehicle-to-infrastructure (V2I) communication	V2I communication allows vehicles to exchange information with traffic infrastructure, such as traffic lights and road signs. AI can process this data to enhance traffic signal coordination, provide real-time updates to drivers, and improve safety by alerting vehicles to potential hazards.
Autonomous Vehicle integration	AI facilitates the integration of Autonomous Vehicles into urban traffic systems. By coordinating the movements of Autonomous Vehicles with traditional traffic and infrastructure, AI can help to manage traffic flow more efficiently and safely, reducing the likelihood of accidents and improving overall traffic conditions.
Public transportation optimization	AI can optimize the scheduling and routing of public transportation based on real-time demand and traffic conditions. This ensures that buses, trams, and other public transport services operate more efficiently, reducing waiting times for passengers and improving the overall reliability of the service.
Environmental monitoring and management	AI systems can monitor environmental conditions, such as air quality and noise levels, and adjust traffic management strategies accordingly. By optimizing traffic flow and reducing congestion, these systems can help to minimize the environmental impact of urban transportation.

Source: Authors' own work.

bicycles are avoided. Such systems can employ machine learning-based algorithms in an attempt to detect abnormal activity or unauthorized attempts at accessing the vehicles and trigger alarms or alert the respective authorities. AI can also be combined with user authentication processes such that only duly authorized persons can use or remove the vehicles from the docking stations (Dodia et al., 2023; Aurangzeb et al., 2024).

Additionally, certain authors (Fong et al., 2023) have studied AI incorporation in green micromobility in Smart Cities. Various smart apps, safety apps, and information apps are founded on an end-to-end vehicle-to-infrastructure network and vehicular cloud integration solution. AI, thus, is used on adaptive modulation and self-aware prognosis for the resolution of parking problems and road traffic congestion problems in a green Smart City environment (de Mattos et al., 2024).

Gigantic vehicle-to-infrastructure (V2I) networks are the basis for such smart services and applications. Micromobility gadgets such as scooters and bicycles in the network communicate with urban infrastructure elements such as traffic lights, parking lots, and other Internet of Things (IoT) components. There is a sharing of information in real time through interaction that forms the crux for efficient functioning of micromobility services. For instance, real-time traffic data can be employed for route optimization, reducing travel time and energy consumption among users of micromobility. V2I equipment can also be utilized to offer dynamic traffic management, which adjusts traffic signal timing according to the number of micromobility vehicles in a pod to try to enhance traffic efficiency and safety (Carr et al., 2019; Fong et al., 2023).

Integrated solutions based on vehicle cloud-based systems also encourage green micromobility operations. By cloud computing, all the vast amounts of data gathered by the micromobility vehicles and infrastructure can be processed and analyzed in real-time. The maintenance of vehicles and infrastructure becomes feasible through data technology, maintaining the assets in optimal condition and reducing downtime. In addition, cloud computing facilitates the deployment and use of sophisticated capabilities of AI, which can predict and avoid future safety threats, provide advisory assistance in real-time, and provide custom mobility advice to consumers (Karlı and Çelikyay, 2021; Fong et al., 2023).

Safety-focused services based on AI are particularly important for sustainable micromobility. These services can include real-time tracking of vehicle conditions, alerts for anomalies that may lead to mechanical failures, and notifications to users about potential dangers on the road. AI has the potential to get users safer via complex features such as collision warning and avoidance systems, where collisions can be evaded by automated speeding down of micromobility vehicle speed and direction when danger is sensed (Dodia et al., 2023).

Car cloud-based information services and AI give the user required information and updates. These services can give current information regarding the availability of micromobility vehicles, station status, traffic, and weather conditions such as air quality. By giving required and timely information, these services facilitate the user to make the right decision about their journey, making their micromobility journey more efficient and convenient (Fong et al., 2023).

5.3 FUTURE OF URBAN MOBILITY WITH AI

Smart mobility of the future holds promise for drastic change with innovation in green practices, connectivity, and AI. As cities expand and urban populations grow, sustainable, efficient, and reliable transport systems become more critical. Smart mobility of the future will transform how we move our cities, making transport efficient, sustainable, and available to all. Due to AI and advanced technologies, we are able to produce smarter, more integrated city environments that maximize the well-being of all residents.

The first that must be mentioned are AV, not cars but trams, buses, and trains as well. Due to advanced machine learning and AI algorithms, AVs read and react to environments in real-time. The technology minimizes human mistake, the largest cause of accidents, and thus greatly improves road safety. The AVs will also enhance mobility of disabled and elderly persons, providing them more freedom of movement and choice. In addition, with the evolving AV technology, we can expect smoother traffic and fewer bottlenecks because AVs can talk to and negotiate with one another, adjusting their routes and velocities to optimal speeds. There already exists subsystems on vehicles that assist the drivers, but this technology will be upgraded to autonomize the car. On the other hand, AVs promise transformative transport but raise several key issues. Safety tops the bill, including issues of liability if something goes wrong and the robustness of AV technology under various conditions. Ethical concerns include decision-making computer programs in times of emergency and the privacy ramifications of vast amounts of data collection. There is a requirement for regulatory policies to ensure that AVs will follow traffic rules and drive safely over several jurisdictions. Ready infrastructure is vital, involving massive investment in upgrading for AV performance accommodation. Public acceptance depends on confidence in AV safety and reduction of societal adaptability hurdles. Providing equal access to the advantages of AVs and integrating them into conventional transport systems are also key challenges.

Smart mobility can look forward to converging various modes of transport, such as public transport, rideshare, bike-share, and micro-modes such as scooters. AI-driven platforms will play a central role in optimizing route planning, offering real-time updates, and allowing smooth transitions between different modes of transport. All this will introduce the convenience of switching from bus to bike-share or from train to scooter, thereby enhancing the ease and efficiency of urban transport.

Secondly, electric cars can prove to be the drivers of green urban transport in the future. AI technologies will most probably optimize EV charging infrastructure, overseeing energy distribution so as to provide efficient and guaranteed access to charging points. Enhanced battery technology will improve driving ranges and lower charging times, yet further contribute to making EVs user-friendly on a daily basis. Integrating renewable power sources with EV charging infrastructure will also become maximally sustainable. Urban mobility's carbon footprint will be lessened. More sophisticated AI algorithms will also be vital for traffic management, by monitoring real-time sensor, camera, and connected-device data to regulate traffic flow. Such systems can adjust the timing of traffic lights dynamically, provide detour information, and enforce variable toll regimes to vacate roads. Predictive analytics will allow traffic management systems to forecast and avoid congestion bottlenecks, improving road safety overall and lessening travel time.

Shared mobility modes like car-sharing, ride-hailing, and bike-sharing can become ubiquitous in the coming years. AI can improve such services by streamlining fleet management, forecasting patterns of demand, and placing vehicles where there is high demand. Shared mobility will thus improve in terms of efficiency, reliability, and convenience for customers, offering them flexible and low-cost alternatives to car ownership.

In addition, travel will be given by AI apps in a personalized manner, optimizing the modes of transport according to preferences, usage patterns, and real-time status. Personalized mobility experiences through optimized usage will increase satisfaction and foster intelligent mobility service usage. For instance, a location-enabled AI app can recommend the best route back home considering traffic, weather conditions, and the user's historical travel profiles. It can make people live more comfortably in Smart Cities.

The future of Smart Cities and AI in the area of infrastructure can include smart infrastructure with the capability to cater to autonomous and connected vehicles.

Smart roads with sensors and communication networks will enable vehicle-to-infrastructure management, making the transportation system more efficient and secure. AVs and connected cars will have to be taken into account by urban planners while designing them, keeping in mind such needs as smart traffic signals and special lanes to fit seamlessly into the existing transportation infrastructure. There are huge challenges regarding urban development, construction, and investment in infrastructure here. For Smart Cities, safety and security are extremely important.

New risks and challenges arise with technological advancement. For this reason, AI must be used in order to increase the security and safety of citizens. This can be accomplished through surveillance and quick response to issues and crimes. Now, there is also frenzy about smart cameras that report services about unpleasant incidents like attacks. Also, AI will have to be harnessed to improve car safety features and transport systems. Predictive maintenance will allow the identification and correction of potential faults at an early stage, lowering the risk of breakdowns and accidents. Real-time hazard detection systems will warn drivers and pedestrians about dangers, to supplement road safety. Also, AI will have a vital part in handling emergency response, giving first responders real-time feedback to enhance their response time and efficiency. Another significant area of concern is emissions. Minimizing the environmental impact of transport will be a focus area in the future of smart mobility. AI will aid emissions reduction by optimizing energy efficiency, facilitating low-emission and zero-emission vehicles, and inducing the adoption of cleaner transport modes. For example, AI can optimize routing and scheduling of electric buses to ensure that they travel the most energy-efficient routes and schedules and make a greater contribution to cleaner and greener cities. Lastly, the moral issue has to be addressed.

AI can become very troublesome with regard to the law.

With smart mobility technologies developing further and further, regulatory strategies and ethics are the most crucial.

Policymakers are going to have to deal with issues of data privacy, cybersecurity and nondiscrimination to make sure that the gains from smart mobility are distributed fairly across society. This is a highly intricate, imprecise, and troublesome issue. Laws, for instance, will have to provide for the safe functioning of antivirus software and safeguarding and ethical utilization of information gathered by AI systems. Access to secure data storage technologies and methods of encryption is pivotal toward avoiding unauthorized access to data and confidentiality. Legal processes have to be instituted in cases of data theft or tampering with data. Preventive strategies against such acts need to be developed as well.

Table 5.3 summarizes the future of AI in Smart Cities in the field of smart mobility. The table consists of columns showing the area and then a description of a given problem or solution.

By embracing these advancements, the future of smart mobility promises to create more efficient, sustainable, and accessible urban transportation systems, significantly improving the quality of life for city residents. The integration of AI and other innovative

TABLE 5.3 Future Directions of AI in Smart Mobility

Area	Description
Autonomous Vehicles	Autonomous Vehicles are expected to become more prevalent, leveraging AI for safe and efficient navigation. These self-driving cars will reduce human error, decrease accidents, and provide greater accessibility to transportation, particularly for individuals with disabilities and the elderly.
Integration of multi-modal transportation	Future smart mobility will emphasize seamless integration of various transportation modes, including public transit, ridesharing, bike-sharing, and micromobility options like scooters. AI-driven platforms will optimize route planning and provide real-time updates, making it easier for users to switch between different modes of transport.
Electric and sustainable vehicles	The shift toward electric vehicles will continue to gain momentum, supported by AI technologies that optimize charging infrastructure and manage energy distribution. Innovations in battery technology and renewable energy integration will further enhance the sustainability of urban transportation.
Smart traffic management	Advanced AI algorithms will play a crucial role in managing traffic flow, reducing congestion, and improving overall transportation efficiency. Real-time data from sensors and connected devices will enable dynamic traffic signal adjustments and predictive analytics to prevent bottlenecks and enhance road safety.
Shared mobility services	The popularity of shared mobility services, such as car-sharing, ride-hailing, and bike-sharing, will increase. AI will enhance these services by optimizing fleet management, predicting demand patterns, and ensuring the availability of vehicles in high-demand areas.
Personalized mobility experiences	AI-driven applications will offer personalized travel assistance, tailoring transportation options based on individual preferences, habits, and real-time conditions. This will improve user experience and encourage the adoption of smart mobility solutions.
Infrastructure and urban planning	Future urban planning will incorporate smart infrastructure designed to support autonomous and connected vehicles. Smart roads, equipped with sensors and communication technologies, will facilitate vehicle-to-infrastructure (V2I) communication, enhancing the overall efficiency and safety of transportation networks.
Enhanced safety and security	AI will continue to advance safety features in vehicles and transportation systems. Predictive maintenance, real-time hazard detection, and emergency response coordination will contribute to safer travel environments.
Sustainability and environmental impact	Reducing the environmental footprint of transportation will remain a priority. AI will assist in minimizing emissions, optimizing energy use, and promoting the adoption of low-emission and zero-emission vehicles, contributing to cleaner and greener cities.
Regulatory and ethical considerations	As smart mobility technologies evolve, regulatory frameworks and ethical considerations will play a significant role. Policymakers will need to address issues related to data privacy, cybersecurity, and equitable access to ensure that the benefits of smart mobility are distributed fairly across all segments of society.

Source: Authors' own work.

technologies will play a critical role in realizing this vision, driving the evolution of transportation into a more connected and intelligent future.

BIBLIOGRAPHY

Abbas, Q., Ahmad, G., Alyas, T., Alghamdi, T., Alsaawy, Y., and Alzahrani, A. (2023). Revolutionizing urban mobility: IoT-enhanced autonomous parking solutions with transfer learning for smart cities. *Sensors*, 23(21), 8753.

Ahmed, M., Zheng, Y., Amine, A., Fathiannasab, H., and Chen, Z. (2021). The role of artificial intelligence in the mass adoption of electric vehicles. *Joule*, 5(9), 2296–2322.

Alanazi, F. (2023). Development of smart mobility infrastructure in Saudi Arabia: A benchmarking approach. *Sustainability*, 15(4), 3158.

Almatar, K.M. (2024). Implementing AI-driven traffic signal systems for enhanced traffic management in Dammam. *International Journal of Sustainable Development & Planning*, 19(2), 781–790.

Amir, M., and Haque, A. (2024). Deep learning-based autonomous vehicle to vehicle detection for smart traffic monitoring in smart cities. In: A. Haque, A. Kalam, & H. Sharma (Eds.), *Smart Cities: Power Electronics, Renewable Energy, and Internet of Things*. Boca Raton, FL: CRC Press, pp. 157–176.

Aurangzeb, S., Aleem, M., Khan, M. T., Anwar, H., and Siddique, M.S. (2024). Cybersecurity for autonomous vehicles against malware attacks in smart-cities. *Cluster Computing*, 27(3), 3363–3378.

Bajaj, D.K., and Siddharth, P. (2023). Artificial Intelligence (AI) in electrical vehicles. In *Recent Advances in Energy Harvesting Technologies*. River Publishers, pp. 57–76.

Belaïd, F., Amine, R., and Massie, C. (2023), Smart Cities Initiatives and Perspectives in the MENA Region and Saudi Arabia. In: J.M. Díaz-Sarachaga, & J.C. Álvarez-Antón (Eds.), *Smart Cities: Social and Environmental Challenges and Opportunities for Local Authorities*. Cham, Switzerland: Springer, pp. 295–313.

Bokolo, A.J. (2023). Inclusive and safe mobility needs of senior citizens: implications for age-friendly cities and communities. *Urban Science*, 7(4), 103.

Boru İpek, A. (2023). Multi-objective simulation optimization integrated with analytic hierarchy process and technique for order preference by similarity to ideal solution for pollution routing problem. *Transportation Research Record*, 2677(1), 1658–1674.

Cao, Y., Cui, J., Liu, S., Li, X., Zhou, Q., and Hu, C. (2023). A holistic review on e-mobility service optimization: Challenges, recent progress and future directions. *IEEE Transactions on Transportation Electrification*, 10(2), 3712–3741

Carr, K., Rojas, J., and Chen, X. (2019). Autonomous refueling strategies using vehicle-to-infrastructure communication in smart cities. In *2019 IEEE 16th International Conference on Mobile Ad Hoc and Sensor Systems Workshops (MASSW)*. IEEE, pp. 77–81.

Chen, Q., and Folly, K.A. (2022). Application of artificial intelligence for EV charging and discharging scheduling and dynamic pricing: A review. *Energies*, 16(1), 146.

Dharavat, N., Golla, N.K., Sudabattula, S.K., Velamuri, S., Kantipudi, M.V.V.P., Kotb, H., and AboRas, K.M. (2023). Impact of plug-in electric vehicles on grid integration with distributed energy resources: A review. *Frontiers in Energy Research*, 10, 1099890.

Dodia, A., Jain, A., Saini, V., and Kantipudi Prasad, M.V.V. (2023). Advanced connected autonomous vehicle (ACAV) with security and system design aspects for intelligent smart cities.

7th IET Smart Cities Symposium (SCS 2023), Hybrid Conference, Bahrain, 2023, pp. 257–261, doi: 10.1049/icp.2024.0934

Duggal, A.S., Singh, R., Gehlot, A., Gupta, L.R., Akram, S.V., Prakash, C., Singh, S., and Kumar, R. (2021). Infrastructure, mobility and safety 4.0: Modernization in road transportation. *Technology in Society*, 67, 101791. https://doi.org/10.1016/j.techsoc.2021.101791

Elhusseini, H., Assi, C., Moussa, B., Attallah, R., and Ghrayeb, A. (2020). Blockchain, AI and smart grids: The three musketeers to a decentralized EV charging infrastructure. *IEEE Internet of Things Magazine*, 3(2), 24–29.

Fong, B., Kim, H., Fong, A.C.M., Hong, G.Y., and Tsang, K.F. (2023). Reliability optimization in the design and implementation of 6G vehicle-to-infrastructure systems for emergency management in a smart city environment. *IEEE Communications Magazine*, 61(8), 148–153

Fong, B., Fong, A.C.M., and Hong, G.Y. (2023). Sustainable micromobility management in Smart Cities. *IEEE Transactions on Intelligent Transportation Systems*, 10(2), 15890–15896. https://doi.org/10.1109/TITS.2023.3292377

Ghaffarpasand, O., Jahromi, A.M., Maleki, R., Karbassiyazdi, E., and Blake, R. (2022). Intelligent geo-sensing for moving toward smart, resilient, low emission, and less carbon transport. In: M. Asadnia, A. Razmjou, & A. Beheshti (Eds.), *Artificial Intelligence and Data Science in Environmental Sensing*. Amsterdam, Netherlands: Elsevier, pp. 39–55.

Kamran, M., Ashraf, S., Salamat, N., Naeem, M., and Hameed, M.S. (2023). Smart city design plan selection through single-valued neutrosophic probabilistic hesitant fuzzy rough aggregation information. *Journal of Intelligent & Fuzzy Systems*, 45(6), 1–45. https://doi.org/10.3233/JIFS-224

Karlı, R.G.Ö., and Çelikyay, S. (2021). Current trends in smart cities: Shared micromobility. In: M.B. Ahmed, A.A. Boudhir, & A. Boukhris (Eds.), *The Proceedings of the International Conference on Smart City Applications*. Cham: Switzerland: Springer, pp. 187–198.

Kubik, A. (2023). The use of artificial intelligence in the assessment of user routes in shared mobility systems in smart cities. *Smart Cities*, 6(4), 1858–1878.

Kumar, P., Hariharan, K., and Manikandan, M.S.K. (2023). Hybrid long short-term memory deep learning model and Dijkstra's Algorithm for fastest travel route recommendation considering eco-routing factors. *Transportation Letters*, 15(8), 926–940.

Leali, F., and Pasquale, F. (2022). The living lab for autonomous driving as applied research of MaaS models in the smart city: The case study of MASA—Modena Automotive Smart Area. In: E. Arbizzani, E. Cangelli, C. Clemente, F. Cumo, F. Giofrè, A.M. Giovenale, M. Palme, & S. Paris (Eds.), *International Conference on Technological Imagination in the Green and Digital Transition*. Cham: Switzerland: Springer, pp. 273–284.

Lee, H., Chatterjee, I., and Cho, G. (2023). A systematic review of computer vision and AI in parking space allocation in a seaport. *Applied Sciences*, 13(18), 10254.

Li, C., Dong, Z., Chen, G., Zhou, B., Zhang, J., and Yu, X. (2021). Data-driven planning of electric vehicle charging infrastructure: A case study of Sydney, Australia. *IEEE Transactions on Smart Grid*, 12(4), 3289–3304.

Louati, A., Louati, H., Kariri, E., Farahat, M. A., and El-Hoseny, H.M. (2024). Sustainable smart cities through multi-agent reinforcement learning-based cooperative autonomous vehicles. *Sustainability*, 16(5), 1779.

Ma, S., Wang, S., Ma, Z., and Qi, Z. (2024). Internet of autonomous vehicles for the distribution system of smart cities. *SAE Technical Paper*.

Machado, C.A.S. De Salles Hue, N.P.M., Berssaneti, F.T., and Quintanilha, J.A. (2018). An overview of shared mobility. *Sustainability*, 10(12), 4342. https://doi.org/10.3390/su10124342

Munhoz, P.A.M.S.A., Dias, F.D.C., Chinelli, C.K., E Silva, W.D.S., and Soares, C.A.P. (2020). Smart mobility: The main drivers for increasing the intelligence of urban mobility. *Sustainability*, 12(24), 10675.

Al Mamlook, R.E., Zahrawi, M., Gharaibeh, H., Nasayreh, A., and Shresth, S. (2023). Smart traffic control system for Dubai: A simulation study using YOLO Algorithms. In *2023 IEEE International Conference on Electro Information Technology (eIT)*. IEEE, pp. 254–264.

Mandžuka, S. (2019). Cooperative systems in traffic technology and transport. In *New Technologies, Development and Application 4*. Springer, pp. 299–308.

Manfletti, C., Guimarães, M., and Soares, C. (2023). AI for space traffic management. *Journal of Space Safety Engineering*, 10(4), 495–504.

Maniyar, K.N., Verma, J.P., and Sharma, N. (2024). Bike sharing systems: A green IT application for smart cities. In *Green Computing for Sustainable Smart Cities*. CRC Press, pp. 150–171.

Mattos, E.P.D., Domingues, A.C.S.A., Silva, F.A., Ramos, H.S., and Loureiro, A.A.F. (2024). Protect your data and I'll rank its utility: A framework for utility analysis of anonymized mobility data for smart city applications. *Ad Hoc Networks*, 163, 103567.

McKinsey Global Institute. (2018). *Smart Cities: Digital Solutions for a More Livable Future*. McKinsey Company.

Mosayebi, M., Gheisarnejad, M., Farsizadeh, H., Andresen, B., and Khooban, M.H. (2022). Smart extreme fast portable charger for electric vehicles-based artificial intelligence. *IEEE Transactions on Circuits and Systems II: Express Briefs*, 70(2), 586–590.

Moumen, I., Abouchabaka, J., and Rafalia, N. (2023). Enhancing urban mobility: Integration of IoT road traffic data and artificial intelligence in smart city environment. *Indonesian Journal of Electrical Engineering and Computer Science*, 32(2), 985–993.

Musa, A.A., Malami, S.I., Alanazi, F., Alshammari, M., and Haruna, S.I. (2023). Sustainable traffic management for smart cities using Internet-of-Things-Oriented Intelligent Transportation Systems (ITS): Challenges and recommendations. *Sustainability*, 15(13), 9859.

Nousias, S., Tselios, C., Bitzas, D., Moustakas, K., and Chatzigiannakis, I. (2019). Exploiting gamification to improve eco-driving behaviour: The GamECAR approach. *Electronic Notes in Theoretical Computer Science*, 343, 103–116.

Oksman, V. (2018). Smart mobility services and senior citizens: A framework for co-creation and analysing user needs. In *25th ITS World Congress, Copenhagen, Denmark*, p. 10.

Paret, P., Finegan, D., and Narumanchi, S. (2023). Artificial Intelligence for power electronics in electric vehicles: Challenges and opportunities. *Journal of Electronic Packaging*, 145(3), 34501.

Patel, A.R., Roscia, M., and Vucinic, D. (2023). Legal implications for autonomous vehicles mobility in future smart cities. In *2023 IEEE International Smart Cities Conference (ISC2)*. IEEE, pp. 1–5.

Piccialli, F., Giampaolo, F., Prezioso, E., Crisci, D., and Cuomo, S. (2021). Predictive analytics for smart parking: A deep learning approach in forecasting of IoT data. *ACM Transactions on Internet Technology (TOIT)*, 21(3), 1–21.

Raza, A., Baloch, M.H., Ali, I., Hassan, M., and Karim, A. (2022). Artificial Intelligence and IoT-based autonomous hybrid electric vehicle with self-charging infrastructure. In *2022 International Conference on Emerging Technologies in Electronics, Computing and Communication (ICETECC)*. IEEE, pp. 1–6.

Rosa, L., Silva, F., and Analide, C. (2022). Explainable artificial intelligence on smart human mobility: A comparative study approach. In *International Symposium on Distributed Computing and Artificial Intelligence*. Springer, pp. 91–101.

Saini, V., Jain, A., Dodia, A., and Kantipudi Prasad, M.V.V. (2023). Approach of an advanced autonomous vehicle with data optimization and cybersecurity for enhancing vehicle's capabilities and functionality for smart cities. IET Conference Proceedings. https://doi.org/10.1049/icp.2024.0930

Santos, G. (2018). Sustainability and shared mobility models. *Sustainability*, 10(9), 3194.

Sawhney, N. (2023). Contestations in urban mobility: Rights, risks, and responsibilities for Urban AI. *AI & Society*, 38(3), 1083–1098.

Seng, K.P., Ang, L.-M., Ngharamike, E., and Peter, E. (2023). Ridesharing and crowdsourcing for smart cities: Technologies, paradigms and use cases. *IEEE Access*, 11, 18038–18081.

Silva, C., and Martins, F. (2020). Traffic flow prediction using public transport and weather data: A medium sized city case study. In *Trends and Innovations in Information Systems and Technologies: Volume* 2 8. Springer, pp. 381–390.

Subramanian, M., Cho, J., Veerappampalayam Easwaramoorthy, S., Murugesan, A., and Chinnasamy, R. (2023). Enhancing sustainable transportation: AI-driven bike demand forecasting in smart cities. *Sustainability*, 15(18), 13840.

Sun, D., Ou, Q., Yao, X., Ma, W., and Li, W. (2020). Integrated human-machine intelligence for EV charging prediction in 5G smart grid. *EURASIP Journal on Wireless Communications and Networking*, 2020(1), 1–15.

Swarnkar, R., Harikrishnan, R., Thakur, P., and Singh, G. (2022). Electric vehicle lithium-ion battery ageing analysis under dynamic condition: A machine learning approach. *SAIEE Africa Research Journal*, 114(1), 4–13.

Tarawneh, M., AlZyoud, F., and Sharrab, Y. (2023). Artificial intelligence traffic analysis framework for smart cities. In *Science and Information Conference*. Springer, pp. 699–711.

Tekouabou, S.C.K. (2021). Intelligent management of bike sharing in smart cities using machine learning and Internet of Things. *Sustainable Cities and Society*, 67, 102702.

Tselios, C., Nousias, S., Bitzas, D., Moustakas, K., and Chatzigiannakis, I. (2019). Enhancing an eco-driving gamification platform through wearable and vehicle sensor data integration. In *Ambient Intelligence: 15th European Conference, AmI 2019, Rome, Italy, November 13–15, 2019, Proceedings 15*. Springer, pp. 344–349.

Turoń, K. (2023). Factors affecting car-sharing services. *Smart Cities*, 6(2), 1185–1201.

Turoń, K., and Tóth, J. (2023). Innovations in shared mobility—review of scientific works. *Smart Cities*, 6(3), 1545–1559.

Vechione, M., Marrufo, C., Vargas-Acosta, R.A., Nunez-Mchiri, G.G., and Chow, J.Y.J. (2018). Smart mobility for seniors: Challenges and solutions in El Paso, TX, and New York, NY. In *2018 IEEE International Smart Cities Conference (ISC2)*. IEEE, pp. 1–8.

Wieczorek, A., Stecuła, K., and Grebski, W.W. (2024). Methods and techniques supporting energy and media savings in maintenance of public transport buses—state of the art and recommendations. *Energies*, 17(9), 2051.

Wolniak, R. (2023a). Analysis of the bicycle roads system as an element of a smart mobility on the example of Poland provinces. *Smart Cities*, 6(1), 368–391.

Wolniak, R. (2023b). European Union smart mobility–aspects connected with bike road system's extension and dissemination. *Smart Cities*, 6(2), 1009–1042.

Wolniak, R. (2023c). Smart biking w smart sity', *Management & Quality/Zarządzanie i Jakość*, 5(2), 313–328.

Wolniak, R., and Grebski, W. (2023). Smart biking and traditional biking. *Silesian University of Technology Scientific Papers. Organization and Management Series*, 178, 717–734.

Zhao, J., Xi, X., Na, Q., Kadry, S.N., and Kumar, P.M. (2021). The technological innovation of hybrid and plug-in electric vehicles for environment carbon pollution control, *Environmental Impact Assessment Review*, 86, 106506.

Zhou, P., Wang, C., and Yang, Y. (2021). Design and optimization of solar-powered shared electric autonomous vehicle system for smart cities. *IEEE Transactions on Mobile Computing*, 22(4), 2053–2068.

CHAPTER 6

AI in Smart Living

6.1 SMART HEALTHCARE SYSTEMS

One of the important aspects related to smart life in Smart Cities is healthcare. Access to healthcare and trouble-free use of it is very important for every person. People need access to various services in this area at different stages of their lives. Unfortunately, in many countries and areas of the world, access to basic healthcare services is difficult. In the face of an aging population, an increase in the number of chronic diseases and rising treatment costs, healthcare systems around the world are facing enormous challenges. New, more effective, and accessible solutions are needed to improve the quality of services provided and make it easier for patients to contact healthcare. In this context, the term smart healthcare becomes crucial. It means to use modern technologies—such as the Internet of Things (IoT), Artificial Intelligence (AI), big data, or mobile devices—for activities such as health monitoring, diagnosing, treating, and managing medical data. Smart healthcare allows not only faster and more precise actions by doctors, but also greater patient independence and health prevention in everyday life. In Smart Cities, the aim is to ensure that healthcare is not only readily available and free from long waiting queues, but also tailored to the individual needs of people.

The first practical usage is using wearable sensors, such as:

- smartwatches—they can measure heart rate, blood oxygen level, ECG, stress level, sleep, and physical activity and remind you to hydrate (Lu et al., 2016),
- smart bands and smart bracelets—they can monitor steps, sleep, heart rate, and physical activity time (Choi and Kim, 2017; Aissa et al., 2025),

DOI: 10.1201/9781003587613-7

- smart T-shirts or clothing with sensors—they can track breathing, ECG, muscle movements, and posture (ten Bhömer et al., 2025),
- specific smart wearable diabetic healthcare devices—they can continuously measure blood sugar levels, especially useful for diabetics (Kim et al., 2024),
- bio-patch—which is a thin patch for the skin that measures heart rate, breathing, body temperature, and activity level (Kong et al., 2025),
- smart rings—they can monitor sleep quality, cardiovascular parameters, and oxygen saturation (Fiore et al., 2024),
- smart glasses—for different purposes, such as improving eye care for glaucoma patients, e.g., recording intraocular pressure (Albar, 2024); assisting seniors with dementia (Prasad et al., 2025); assisting people visually impaired and blind (Varshney et al., 2025); and helping nurses with their jobs (Challa et al., 2024).

Wearable smart sensors work by collecting biological data using built-in sensors. The collected data is sent to dedicated mobile applications, where the user (or doctor) can analyze it in real time (Mukhopadhyay et al., 2022). This allows for ongoing health monitoring, early detection of irregularities, and personalization of medical care.

Furthermore, smart healthcare methods and devices let continuous health monitoring, and even early disease detection. It is also possible to create personalized treatment plans. AI can analyze big amounts of healthcare data. This enables identifying patterns and optimizing resource allocation within healthcare systems. There are also telemedicine platforms that enable remote medical consultations and diagnostics. They can reduce the need for in-person visits and improve access to healthcare services, especially in underserved urban areas. There are numerous publications in the literature on this aspect of smart living. A. M. Ali et al. (2023) in their paper proposed a method to improve the efficiency of healthcare services in Smart Cities using hybrid optimization techniques. Zhong and Deng (2023), using cloud and IoT, developed a workload-aware healthcare system. They applied an ant colony optimization algorithm. Hassan et al. (2023) presented the principles of an AI-based health monitoring system for Smart Cities. Abirami and Karthikeyan (2023) proposed a digital twin-based healthcare system that offers intelligent virtual care services that support early disease detection and patient-centric diagnosis, even from remote locations.

Another interesting area of development of smart healthcare is systems supporting the elderly and dependent. For example, smart homes can be equipped with motion sensors that monitor the daily activity of residents and automatically notify family or medical services if unusual behavior is detected (e.g., lack of movement for a long time) (Skubic et al., 2009). There are also systems that remind patients to take medication—these can be smart dispensers or mobile applications that not only inform the patient but also provide feedback to the doctor or caregiver. Thanks to such solutions, it is possible to extend the independence of the elderly and improve their safety in the home environment (Nath and Thapliyal, 2021).

Hospital and medical facility management systems are also developing, which integrate patient data, enable better planning of procedures and staff management. Thanks to the use of AI, it is possible to forecast the occupancy of wards, predict queues for specialists, and dynamically manage medical resources. For example, some hospitals use systems for automatic analysis of imaging test results (e.g., X-ray, MRI), supporting radiologists in diagnosing cancers or degenerative changes. In practice, this translates into shorter diagnosis times, reduced staff workload, and improved treatment quality (Atta and Talamo, 2020).

It is also worth mentioning the use of chatbots and medical assistants based on AI (Alam et al., 2022). They allow patients to quickly obtain answers to health questions, help interpret test results, and schedule appointments. These solutions reduce the workload on medical hotlines and registrations, while offering greater access to information for patients—24 hours a day. Data from various sources—health applications, sensors, and medical records—is increasingly being integrated, creating a single, coherent patient health management system, which is consistent with the idea of integrated and personalized medical care in Smart Cities (Zhou et al., 2020).

Another intelligent healthcare development trend is the use of predictive analytics and AI for the prevention of chronic diseases in their earlier stages and the management of healthcare in cities (Nwankwo et al., 2024). AI then performs the analysis of bulk electronic health record data, mobiles, environment, and behavior patterns. AI algorithms can identify persons with higher chances of developing chronic conditions (e.g., cardiovascular disease, diabetes, or respiratory conditions). This makes possible the creation of lifestyle recommendations, tailored prevention plans, and screening. AI is also useful in pandemics and infectious disease outbreaks. Smarter systems can be utilized to enable real-time epidemiological monitoring, contact tracing, and efficient allocation of resources, maximizing the impact of public health interventions (Gendy and Yuce, 2022).

In summary, smart healthcare is a highly significant part of a Smart City. It enables more effective, cost-effective, and patient-centric medical services. Owing to the aging global population, growing prevalence of chronic conditions, and rising healthcare costs, the implementation of smart technologies in healthcare systems is necessary. The future of medicine will rest primarily on data-driven decisions, real-time tracking, and personalized care made possible by AI and IoT solutions.

6.2 INTELLIGENT HOME AUTOMATION

Many solutions within smart living concern the home infrastructure. They talk about smart homes, as well as smart systems that are used in homes or other buildings. These systems allow for remote and automated operation. They are often connected to an external operating system. They store and process huge amounts of data. All this is to make people's lives more and more comfortable and adaptable to their needs and individual preferences.

There are many such solutions in the home in literature and practice. For example, in the publication by Chatwattana et al. you can read about intelligent lighting systems, ventilation systems, closing and opening curtains (Chatwattana et al., 2024). An intelligent lighting system automatically adjusts the light intensity in a room depending on various factors, such as the time of day, the presence of household members, and the amount of natural light. By using sensors and AI algorithms, not only is user comfort increased, but energy is also saved. This can translate into reduced lighting costs. In terms of smart living, this is a good solution for users who may, for example, have lower light levels in the evening when they go to bed. In the morning, natural sunlight can be let in instead of using artificial lighting. AI also allows learning user habits and preferences to apply the expected approach not only for a given home but also for a given user. Another solution—a ventilation system—not only monitors the air temperature in the house but can also change and adjust the temperature and airflow, turn the ventilation on and off at a specified desired time, and operate according to a preset schedule.. However, such a system also has other functions: it can also take care of the air quality in the house or the humidity level. Sensor systems can detect, for example, the presence of carbon dioxide (CO_2), particulate matter ($PM_{2.5}$), smoke, dust, or outdoor pollution. When the air quality deteriorates, the system can, for example, turn on the ventilation, air filter, or open a window. When it comes to humidity levels, too low humidity can cause dry skin and respiratory tracts, while too high humidity promotes mold and poor well-being. An intelligent system can turn on a humidifier or ventilation depending on what is needed. In addition, it can also work with heating or cooling systems to ensure optimal indoor conditions and energy savings. Automatic curtains are also an interesting solution. They operate based on schedules, weather conditions or the presence of residents. Thanks to AI support, they can change their angle and window coverage depending on human preferences, looking for the optimal solution also in the energy saving category. Such curtains can work with other devices—e.g., heating—and choose the best scenario in terms of letting in natural light, which also provides additional heat energy. The system can, for example, cover windows on a hot day to prevent rooms from overheating, or open them when it detects the presence of a user, letting in daylight. More research on the optimization of blinds and curtains in smart homes can be found in the literature, e.g., Karmakar et al. (2017) and Souza et al. (2018). Some of them also presents special applications for users (Havran et al., 2025; Tian, 2022).

Similarly to ventilation in smart homes, there are other solutions related to temperature. These are systems used to heat the home. Considering the overall global energy production, it is worth considering the use of AI-based systems that will help reduce energy consumption. This approach is important for users, as effective solutions can reduce their costs, while also contributing to lowering the environmental impact of energy production. Of course, this is a small percentage—although any pro-ecological activity is important—looking at the overall calculation. Heating systems in a smart home (home automation) are solutions that automatically control the temperature in the apartment. Thanks to sensors and AI, the system itself knows when to increase or decrease the heating—for example, it lowers the temperature when no one is at home and raises it before the household returns. You can also control the heating remotely, e.g., via an application on your phone. Such systems help save energy and provide thermal comfort without the need to manually adjust the radiators. Such solutions help reduce energy consumption, which is confirmed in numerous publications (Schuetz et al., 2020; Okoro et al., 2023).

Another solution used in homes is intelligent security systems. These are solutions that increase the protection of residents and property. They use video surveillance cameras, facial recognition technology (Pawar et al., 2018), and motion sensors (Ding et al., 2011) to quickly detect potential threats. The last one often utilizes a deep learning model for motion recognition and classification based on the detected movement patterns. In case of a disturbing situation, the system can automatically notify the homeowner, e.g., via an application on the smartphone (Taiwo et al., 2022). What is more, there are possible management functions like remote access—for example opening the door for friends or a courier without leaving work. Such solutions provide a sense of security and full control over what is happening at home—even when no one is there. Additionally, smart homes also offer systems for fire prevention and detection (Saeed et al., 2018).

Another important convenience offered by home automation systems is the management of household appliances in a smart home. By connecting household appliances to a Wi-Fi network and mobile applications, users can remotely control the operation of devices such as a washing machine, oven, dishwasher, or refrigerator—regardless of where they are. For example, you can turn on the laundry while still at work or set the oven to be warm by the time you get home. Naturally, the problematic issue is consumer acceptance of such devices (Mert et al., 2008). Additionally, many systems using AI can analyze energy rates and automatically start devices at times when energy is cheapest. This approach can translate into savings for users. Such solutions not only increase the comfort of use but also contribute to effective energy management and care for the environment.

Last but not least, it can be mentioned that voice assistants and voice control systems play an increasingly important role in intelligent home automation (Ali et al., 2017; Venkatraman et al., 2021). They allow users to control various elements of the home—such as lighting, music, curtains, or heating—simply by speaking. Additionally, they can integrate with calendars, provide reminders, deliver weather updates, and much more, making everyday life more convenient and connected. These technologies are becoming more intuitive and user-friendly. Voice assistants create a more natural interaction with smart home systems, removing the need for complex interfaces or manual operation. As their capabilities grow, they are expected to become true digital companions, supporting everyday routines and enhancing the quality of life at home. Who wouldn't want an AI companion to talk to like a chat friend?

Finally, it is worth mentioning that on the one hand, smart devices and home automation solutions are supposed to make life easier for residents and make their living more intelligent. On the other hand, in practice, many problems may arise with the use of such solutions. There may also be errors and unwanted situations. Hence, potential recipients often show resistance or even fear of such solutions. Naturally, many factors motivate consumers' decisions to use smart home services, and these may include: social influence, ostentation of social status (which further divides society in terms of prosperity),

TABLE 6.1 Examples of Benefits, Threats, and User Resistance to Intelligent Home Automation Technology

No.	Benefits	Threats	User Resistance
1.	Increased comfort	Risk of hacking and cyberattacks	Fear of losing privacy
2.	Energy savings and lower utility bills	Dependence on internet and electricity	Lack of knowledge or skills to operate smart technology
3.	Remote control of home (e.g., via smartphone apps)	Compatibility issues between different devices	Fear of complicated installation and high costs
4.	Improved home security (sensors, alarms, smart locks)	System malfunctions—e.g., AI or app failure	Lack of trust toward Artificial Intelligence
5.	Personalized settings (e.g., lighting, temperature)	Risk of personal data leakage due to weak cybersecurity	Concern about becoming too dependent on technology
6.	Support for elderly and people with disabilities	High cost of implementing advanced smart solutions	Perception that it's just a gadget, not a real necessity
7.	Integration with renewable energy systems	Difficult maintenance or lack of technical support	Lack of trust in durability and reliability of the devices

Source: Author's own work.

hedonistic motivation, general attitude, or simply perceived ease of use, as well as the issue of connectivity and expectations regarding performance (Shivdas et al., 2025). However, one of the main barriers is the fear of losing privacy—users are afraid that smart devices will eavesdrop on them or collect too much data about their daily lives. Another problem is the difficulty of use—older or less technical people may have trouble understanding how applications or systems work. There is also a concern about security, for example that hackers can break into the system and take control of devices at home. For some, the cost of implementation is also a barrier—smart systems are often associated with something expensive. Finally, some people do not trust new technologies and prefer traditional, proven solutions. Table 6.1 presents basic examples of benefits, threats, and user resistance to intelligent home automation technology.

6.3 AI-ASSISTED EDUCATION

Smart living is also related to education, which is supported by AI in Smart Cities. AI can support not only students and teachers but also citizens of Smart Cities. AI-Assisted Education is a modern approach to education that uses advanced algorithms to personalize content, optimize the teaching process, and adapt the porcupine of teaching to a given person depending on their level or educational expectations.

One of the main features provided by AI in education is the personalization of learning. Algorithms analyze the student's progress and, thanks to this, can adapt teaching materials to their individual needs—the pace of work, learning style, or level of advancement. Thanks to this, each student can learn at their own pace, with material tailored to their preferences and difficulties. Intelligent educational assistants are also increasingly used—chatbots or virtual teachers available 24/7, who explain incomprehensible issues and answer students' questions in real time. Surely, such an approach can be useful in pandemic situations or when restrictions occur due to another problem. Then direct contact with people is limited and learning with AI is a good solution to maintain the continuity of learning. Various subjects can be studied with AI, but there is currently quite a lot being talked about learning foreign languages with the usage of AI assistants. Such a way of studying aims to improve foreign language learners' experiences through personalized, interactive, and collaborative learning (Semerikov et al., 2021).

However, the rapid development of AI means that it can increasingly better transfer knowledge and adapt to the user. In some areas of education, such as learning languages or programming, it has a good chance of being more precise than a human teacher. Therefore, the following question arises: will a human teacher still be needed in the future? Table 6.2 presents the issues in the area of human teacher and AI assistant in various areas. The table shows that although AI assistants offer quick access to knowledge, personalization, and the ability to teach on a large scale 24/7, they are not able to replace teachers in areas requiring empathy, values, creativity, and interpersonal relationships. The best solution seems to be a hybrid model, combining the advantages of both approaches.

Another important aspect of AI in education is content recommendation systems. Based on the interests, previous results and goals of the student, AI can suggest films, articles, courses or tasks that support the development of specific competences. This is similar to the product recommendations for purchase that often appear on websites—they are personalized based on the user's previous purchases, interests, movements on the pages, etc. AI can also use predictive analysis to forecast potential student difficulties or the risk of failure—which gives the teacher and parents a chance to react quickly and provide support.

AI also relieves teachers of time-consuming tasks—for example, by automatically marking tests or essays. Such systems operate quickly, are objective, and learn from the

TABLE 6.2 Comparison of Human Teachers and AI Educational Assistants in Smart Education

No.	Human Teacher	AI Assistant	Remarks
Relationship	Shows real emotions, offers emotional support	Simulates empathy, lacks real emotional understanding	AI cannot replace genuine human connection, especially in emotional situations
Personalized approach	Uses experience and intuition	Analyzes large datasets to personalize content	AI is faster but lacks full context of a student's life
Assessment and grading	Can be subjective, time-consuming	Fast, consistent grading (mainly for structured tasks)	Creative tasks remain challenging for AI to evaluate fairly
Availability	Limited to school hours and physical presence	Available 24/7	AI enables learning anytime, helps bridge gaps in access
Teaching values and ethics	Models values through behavior and discussion	Lacks moral awareness or value systems	AI cannot truly teach empathy, critical thinking, or ethics
Creativity and inspiration	Can inspire, motivate, and show passion	Limited to pre-existing data, lacks true creativity	AI cannot tell personal stories or adapt spontaneously (do not have human life experience)
Cost and access	More expensive, limited to smaller groups	Cost-effective, can teach thousands of users at the same time	AI is good for mass education support but not as a full replacement for teachers

Source: Author's own work.

analysis of large datasets, which increases their accuracy. E-learning platforms integrated with AI increasingly enable dynamic adjustment of tasks, interface, and learning path to the current progress and needs of the student.

AI is also used in the organization of work in schools and universities. Systems based on AI support the management of class schedules, resource allocation, and cost optimization. It is also an invaluable aid in the education of students with special needs—AI can, for example, automatically read texts aloud, enlarge images, translate into sign language, or adjust interfaces.

Education in Smart Cities can also be done through the use of virtual reality (VR), augmented reality (AR), and mixed reality (MR). These technologies offer many possibilities. Pupils and students can conduct experiments in VR without the need for physical equipment, e.g., chemistry, biology, and physics. Schools can create "digital classrooms" that work even remotely and support students from excluded areas (Das, 2023). AR educational applications display 3D models of cells, planets, engines, or word translations "live" using cameras in smartphones or tablets (Lindner et al., 2019). Another solution is visiting museums in VR, reproducing historical events with the possibility of interaction (Carrozzino and Bergamasco, 2010). Moreover, Smart Cities have developed the idea of the metaverse. In the literature, there can be found some examples on XR technology in IoT applications in Smart Cities (Bibri and Jagatheesaperumal, 2023). For example, interactive virtual events and conferences, virtual markets and digital economies, urban planning and simulations, socio-cultural experiences, or various educational and professional simulations (e.g., for employees). In the future, it is also expected that each student may have a virtual AI "mentor" who adapts the material to their learning style and interests (Zhu et al., 2025). You can also imagine classes based on real-time city data. Students can learn about air pollution, traffic, and energy consumption by analyzing live data visualized in MR. It is also worth mentioning the possibility of training in digital and civic skills. VR and AR can prepare residents to use digital services (e.g., submitting applications, using health portals). Such solutions would be helpful in social education, especially in times when the discrepancy in digital skills is noticeable—and there is growing concern about digital exclusion (Mubarak and Suomi, 2022).

Finally, it is worth noting that AI also supports the idea of lifelong learning (Masrek et al., 2024). This is so important in the context of dynamic changes in the labor market. Intelligent recommendation systems and microlearning platforms enable city dwellers to develop digital, professional, and personal competences in a flexible manner that is made to their needs.

BIBLIOGRAPHY

Abirami, L., and Karthikeyan, J. (2023). Digital twin-based healthcare system (DTHS) for earlier Parkinson disease identification and diagnosis using optimized fuzzy based k-nearest neighbor classifier model. *IEEE Access*, 11, 96661–96672. https://doi.org/10.1109/ACCESS.2023.3312278

Aissa, Y. Ben, Saoudi, M.N., Zerarka, N., and Rechachi, B. (2025). Smart bracelet to monitor hospital patients. *2025 International Symposium on INnovative Informatics of Biskra (ISNIB)*, 1–6.

Alam, M., Khan, E., Alam, A., Siddiqui, F., & Tanweer, S. (2023). AI for sustainable smart city healthcare. *China Petroleum Processing and Petrochemical Technology*, 23, 2245–2258.

Albar, A.A. (2024). The use of smart glasses to enhance eye healthcare for glaucoma patients: A prototype design to remotely measure and record eyes' intraocular pressure (IOP). *International Conference on* World, S4, 79–94.

Ali, A.M., Hegazy, A.-E.F., Dahroug, A., and Hassan, K.M. (2023). A proposed model for enhancing the performance of health care services in smart cities using hybrid optimization techniques. *2023 15th International Conference on Computer Research and Development*, ICCRD 2023, 135–142. https://doi.org/10.1109/ICCRD56364.2023.10080010

Ali, A.T., Eltayeb, E.B., and Abusail, E.A.A. (2017). Voice recognition based smart home control system. *International Journal of Engineering Inventions*, 6(4), 1–5.

Atta, N., and Talamo, C. (2020). Facility management services in smart cities: Trends and perspectives. *New Metropolitan Perspectives: Knowledge Dynamics, Innovation-Driven Policies Towards the Territories' Attractiveness*, Volume 1, 220–230.

Bibri, S.E., and Jagatheesaperumal, S.K. (2023). Harnessing the potential of the metaverse and artificial intelligence for the internet of city things: Cost-effective XReality and synergistic AIoT technologies. In *Smart Cities* (Vol. 6, Issue 5, pp. 2397–2429). https://doi.org/10.3390/smartcities6050109

Carrozzino, M., and Bergamasco, M. (2010). Beyond virtual museums: Experiencing immersive virtual reality in real museums. *Journal of Cultural Heritage*, 11(4), 452–458.

Challa, A.K., Lauterbach, M., Metzler, L., and Wehde, A. (2024). Assessing the Usability of Smart Glasses in Nursing Homes: A User Experience Approach. [preprint]

Chatwattana, P., Yodkla, A., Kruachottikul, P., Kovitanggoon, K., and Chancharoen, R. (2024). Smart living BT. *Proceedings of Ninth International Congress on Information and Communication Technology* (X.-S. Yang, S. Sherratt, N. Dey, and A. Joshi (eds.), pp. 55–70). Springer Nature Singapore.

Choi, S.-H., and Kim, S.-I. (2017). A study on the factors affecting the purchase of healthcare smart bands. *Journal of the Korea Convergence Society*, 8(7), 175–181.

Das, N. (2023). Digital education as an integral part of a smart and intelligent city: A short review. In: A. Choudhury, A. Biswas, & S. Chakraborti (Eds.), *Digital Learning Based Education: Transcending Physical Barriers*. Singapore: Springer Nature, pp. 81–96.

Ding, D., Cooper, R.A., Pasquina, P.F., and Fici-Pasquina, L. (2011). Sensor technology for smart homes. *Maturitas*, 69(2), 131–136.

Fiore, M., Bianconi, A., Sicari, G., Conni, A., Lenzi, J., Tomaiuolo, G., Zito, F., Golinelli, D., and Sanmarchi, F. (2024). The use of smart rings in health monitoring—a meta-analysis. *Applied Sciences*, 14(23), 10778.

Gendy, M.E.G., and Yuce, M.R. (2022). Emerging technologies used in health management and efficiency improvement during different contact tracing phases against COVID-19 pandemic. *IEEE Reviews in Biomedical Engineering*, 16, 38–52.

Hassan, A.H., Sulaiman, R. bin, Abdulgabber, M.A., and Kahtan, H. (2023). Balancing technological advances with user needs: User-centered principles for AI-driven smart city healthcare monitoring. *International Journal of Advanced Computer Science and Applications*, 14(3), 365–376. https://doi.org/10.14569/IJACSA.2023.0140341

Havran, V., Bahlai, M., Kruk, B.-I., and Stasiv, D. (2025). Development of a web-based application for smart curtain control. *Computer Design Systems Theory and Practice*, 7(1), 189–196.

Karmakar, G., Roy, S., Chattopadhyay, G., and Xiao, Z. (2017). Dynamically controlling exterior and interior window coverings through IoT for environmental friendly smart homes. *2017 IEEE International Conference on Mechatronics (ICM)*, 487–491.

Kim, T.Y., De, R., Choi, I., Kim, H., and Hahn, S.K. (2024). Multifunctional nanomaterials for smart wearable diabetic healthcare devices. *Biomaterials*, 310, 122630.

Kong, J., Chu, R., Mao, M., Yu, H., Liu, J., Sun, Y., Ge, X., Jin, Z., Huang, W., and Hu, N. (2025). A 3-in-1 wearable and implantable biopatch for integrated in-body energy harvesting, sensing, and self-powered diabetic wound treatments. *Nano Energy*, 136, 110743.

Lindner, C., Rienow, A., and Jürgens, C. (2019). Augmented reality applications as digital experiments for education–an example in the earth-moon system. *Acta Astronautica*, 161, 66–74.

Lu, T.-C., Fu, C.-M., Ma, M.H.-M., Fang, C.-C., and Turner, A.M. (2016). Healthcare applications of smart watches. *Applied Clinical Informatics*, 7(03), 850–869.

Masrek, M.N., Susantari, T., Mutia, F., Yuwinanto, H.P., and Atmi, R.T. (2024). Enabling education everywhere: How artificial intelligence empowers ubiquitous and lifelong learning. *Environment-Behaviour Proceedings Journal*, 9(SI18), 57–63.

Mert, W., Suschek-Berger, J., and Tritthart, W. (2008). Consumer acceptance of smart appliances. *Smart Domestic Appliances in Sustainable Energy Systems (Smart-A)*, 1–46.

Mubarak, F., and Suomi, R. (2022). Elderly forgotten? Digital exclusion in the information age and the rising grey digital divide. *INQUIRY: The Journal of Health Care Organization, Provision, and Financing*, 59, 00469580221096272.

Mukhopadhyay, S.C., Suryadevara, N.K., and Nag, A. (2022). Wearable sensors for healthcare: Fabrication to application. *Sensors*, 22(14), 5137.

Nath, R.K., and Thapliyal, H. (2021). Machine learning-based anxiety detection in older adults using wristband sensors and context feature. *SN Computer Science*, 2(5), 359.

Nwankwo, E.I., Emeihe, E.V., Ajegbile, M.D., Olaboye, J.A., and Maha, C.C. (2024). Artificial Intelligence in predictive analytics for epidemic outbreaks in rural populations. *International Journal of Life Science Research Archive*, 7(1), 78–94.

Okoro, C.K., Nwaorgu, O.A., Odo, K.O., Daniel, S., Kanu, N.D., and Chikelu, P.C. (2023). Application of artificial intelligent techniques to temperature control in automated home heating. *Journal of Inventive Engineering and Technology (JIET)*, 4(2), 29–42.

Pawar, S., Kithani, V., Ahuja, S., and Sahu, S. (2018). Smart home security using IoT and face recognition. *2018 Fourth International Conference on Computing Communication Control and Automation (ICCUBEA)*, 1–6.

Prasad, R.S., Prasad, J.R., Kulkarni, P.A., Farakte, A.B., and Swami, S.G. (2025). Smart assistive technology for senior citizens. In *Predictive Algorithms for Rehabilitation and Assistive Systems* (pp. 165–190). IGI Global Scientific Publishing.

Saeed, F., Paul, A., Rehman, A., Hong, W.H., and Seo, H. (2018). IoT-based intelligent modeling of smart home environment for fire prevention and safety. *Journal of Sensor and Actuator Networks*, 7(1), 11.

Schuetz, P., Melillo, A., Businger, F., Durrer, R., Frehner, S., Gwerder, D., and Worlitschek, J. (2020). Automated modelling of residential buildings and heating systems based on smart grid monitoring data. *Energy and Buildings*, 229, 110453. https://doi.org/https://doi.org/10.1016/j.enbuild.2020.110453

Semerikov, S.O., Striuk, A.M., and Shalatska, H.M. (2021). AI-assisted language education: Critical review. *Educational Dimension*, 4, 1–7.

Shivdas, A., Soman, S., Nair, S.M., and Das, K.S. (2025). Embracing smart living: Exploring factors influencing the adoption of smart home technology. In *Business Sustainability with Artificial Intelligence (AI): Challenges and Opportunities: Volume 1* (E. AlDhaen, A. Braganza, A. Hamdan, and W. Chen (eds.); pp. 987–997). Springer Nature Switzerland. https://doi.org/10.1007/978-3-031-71526-6_88

Skubic, M., Alexander, G., Popescu, M., Rantz, M., and Keller, J. (2009). A smart home application to eldercare: Current status and lessons learned. *Technology and Health Care*, 17(3), 183–201.

Souza, L.C., Rodrigues, J.J.P.C., Scarpioni, G.D., Santos, D.A.A., De Albuquerque, V.H.C., and Dhurandher, S.K. (2018). An IoT automated curtain system for smart homes. *2018 International Conference on Advances in Computing, Communications and Informatics (ICACCI)*, 249–253.

Taiwo, O., Ezugwu, A.E., Oyelade, O.N., and Almutairi, M.S. (2022). Enhanced intelligent smart home control and security system based on deep learning model. *Wireless Communications and Mobile Computing*, 2022(1), 9307961.

ten Bhömer, M., Wang, Q., Bai, Z., Jiang, M., Qu, N., and Wu, Z. (2025). From monitoring to proactive healthcare: Cases exploring embodiment and AI in smart textile design. *The Design Journal*, 1–24.

Tian, H. (2022). Research on the application of wireless sensor in the design of curtain automatic adjustment system. *Highlights in Science, Engineering and Technology*, 24, 62–67.

Varshney, A.S., Chougle, M.E., Patel, C.V., and Chauhan, M.D. (2025). Evaluating usability of "the smart vision glasses" for individuals who are visually impaired and totally blind. *Saudi Journal of Ophthalmology*, 10–4103, 1–6.

Venkatraman, S., Overmars, A., and Thong, M. (2021). Smart home automation—use cases of a secure and integrated voice-control system. *Systems*, 9(4), 77.

Zhong, L. and Deng, X. (2023). A cloud and IoT-enabled workload-aware healthcare framework using ant colony optimization algorithm. *International Journal of Advanced Computer Science and Applications*, 14(3), 824–834. https://doi.org/10.14569/IJACSA.2023.0140394

Zhou, R., Zhang, X., Wang, X., Yang, G., Guizani, N., and Du, X. (2020). Efficient and traceable patient health data search system for hospital management in smart cities. *IEEE Internet of Things Journal*, 8(8), 6425–6436.

Zhu, F., Choi, B., and Tishchenko, V. (2025). Understanding Human Learning Performance through AI Mentor Interactions, AMCIS 2025 Proceedings. Paper No. 1397, Montreal Canada.

CHAPTER 7

AI in Smart People

7.1 EMPOWERING CITIZENS THROUGH AI

Despite the smart living area of Smart Cities, there is another area connected directly with humans, which is called "smart people." This refers to initiatives and technologies that help inhabitants acquire specific skills, knowledge, and resources. It primarily focuses on personalized education, digital literacy, community engagement, and fostering the concept of lifelong learning. There are also some features, referring to people, important in Smart Cities. They include humane, self-aware, and co-operative (Streitz et al., 2022). Self-awareness brings a new definition of the word "smartness." It is connected with initiatives and technologies that help people not only to understand their behavior and habits, but also their impact on the environment. People should pay greater attention to mindfulness, responsibility, and proactiveness when it comes to the sustainability of cities and inclusive communities. In other literature source, factors important in Smart Cities for smart people are the following: open data, communities, collective participation, socio-technical engagement, and empowerment (Feher, 2021). Other scholars stress the importance of climate awareness, enhancing smart governance, and supporting social participation and inclusion in Smart Cities (Obracht-Prondzyńska et al., 2022).

And it is precisely Artificial Intelligence (AI) that should help people in the aforementioned issues. However, it should be emphasized that AI should not focus on technology itself in the development of intelligent environments, but on people. People should be put first in Smart Cities, and AI should help citizens. However, in practice, this issue can be problematic, because the boundary between the development of technology itself and its development for people is blurring. Hence, it is important to create and design solutions based on AI with the good of people in mind, and not with the development

DOI: 10.1201/9781003587613-8

of technology itself in mind. Unfortunately, in today's world, the focus is often on the development of technology for development, for marketing, for the so-called rat race among high-tech companies. Competition for the customer is high and unfortunately the customer (recipient) is often treated as an object, not as a person. Technology, including AI, is supposed to be for people. This issue is also discussed in the literature, for example Streitz discusses the concept of "human-technology symbiosis" (Streitz, 2021). The author discusses the concept of humanitarian, social and cooperative hybrid cities. Such cities are supposed to connect people and technology by ensuring a balance between human control and automation, as well as privacy and intelligence. A Smart City must be "self-aware" and "cooperative" toward its citizens.

Modern cities face a huge challenge: how to build a society that is more inclusive, fair, and responsive to the diverse needs of its residents. AI can be helpful by offering tools that support citizens in their daily lives. It can also help with social integration, increase access to information and services, and improve the quality of life of groups at risk of exclusion. AI can support citizens not only by making it easier to use services, but also by really strengthening their position in public and social life.

One of the most important applications of AI in empowering citizens is digital citizen assistants. Thanks to advanced natural language processing technologies, such solutions can provide information about public services, translate legal regulations into a language understandable to every resident, or help in submitting applications and forms. Digital assistants available 24/7 can also support the elderly, migrants, or people with limited mobility in independently navigating the city service system.

AI also supports civic participation. AI systems can analyze hundreds of thousands of opinions of residents submitted through city applications, internet forums, or social media, helping officials understand the real needs of citizens and respond faster. In turn, residents can use consultation chatbots, which allow them to participate in local debates and decision-making processes without the need to physically attend meetings.

It is also worth emphasizing the role of AI in civic education and developing social competences. Intelligent platforms can develop digital, civic, and professional competences of residents, adapting content to their needs and level of knowledge. In this way, AI can not only equalize opportunities, but also be a tool for activating local communities. It can also support the development of the so-called information society (Abdizoitovich, 2025). The information society is characterized by such a social model in which access to information, knowledge, and digital technologies plays a key role, and economic development, education, and everyday life are strongly dependent on data processing and exchange. It is worth emphasizing that AI plays an important role in such a society, as it helps in the analysis of huge amounts of information, automates decision-making processes, and supports citizens in everyday activities—from education, through healthcare, to participation in social and civic life.

AI can also support local civic initiatives and help build a so-called civil society that will be involved in civic affairs. AI can create intelligent consultation platforms that, for example, analyze residents' opinions and identify the most important topics for local communities. Algorithms can group similar demands, making it easier for local authorities to respond to the needs of residents in a transparent and orderly manner. In other words, AI can unite.

AI can also support the activities of non-governmental organizations by automating data collection, analyzing the needs of beneficiaries, or predicting the effects of actions taken. Thanks to this, local leaders and activists gain tools that increase their effectiveness and allow them to have a real impact on social life in the city. In the long term, this can contribute to increasing citizens' trust in institutions and increasing the sense of agency among residents.

AI can also be used to solve the problem of food waste. In supermarkets, AI can help sort and package food that is suitable for consumption (which would otherwise be thrown away). As it can be read in the article of Agrawal and Panchal (2023), instead of throwing away food, it can be distributed: food parcels can be prepared for people in need. This solution introduces a system for managing surplus food.

7.2 ENHANCING DIGITAL LITERACY AND INCLUSIVITY

In the era of digital transformation and dynamic technological development, the ability to use digital tools is becoming essential in everyday life. In Smart Cities, which increasingly rely on data, applications, and systems based on AI, developing residents' digital skills is crucial. Therefore, one of the key elements of Smart City development should be initiatives supporting citizens' digital education.

The foundation of building an information society in a Smart City is the introduction and development of digital education from an early age. Schools and universities can offer curricula covering not only the basics of computer and internet use, as has been the case so far, but also current topics such as cybersecurity, using public e-services, managing online privacy, and programming basics. For example, in some cities, students learn coding as early as elementary school, which develops their logical thinking and prepares them for the informed use of technology in the future (Price and Price-Mohr, 2018). In recent years, issues related to communicating with AI have also become very important. In recent years, issues related to AI communication, especially GPT chat, have also become very important. As it can be read in the article of Božić and Poola, in education, GPT chat can be used for language learning, writing assistance, automated assessment, and personalized learning (Božić and Poola, 2023). Firstly, GPT can be used to create chatbots and virtual language tutors. These help students practice language skills. They simulate real-world conversations and provide feedback (for example, on grammar, pronunciation, and vocabulary). When it comes to improving writing skills, GPT can analyze students' writing style and make corrections and provide feedback on grammatical, punctuation, and spelling errors. Furthermore, GPT can grade written work. Finally, GPT can be used to create personalized educational content. This can recommend educational resources (textbooks, videos) tailored to a given individual's needs. However, it should be emphasized that GPT should be used as a learning support tool, not as a replacement for teachers. GPT chat can be helpful, but there are also many risks associated with using it. Table 7.1 shows the positive and negative aspects of using GPT chat for educational purposes. However, AI in this area will undoubtedly continue to develop, and its skillful and wise use will become an important essential competency in human life.

Following the matter of digital literacy, it is important to mention that certain Smart Cities are investing in local hubs of digital assistance, where seniors and adults can

TABLE 7.1 Positive and Negative Aspects of Using GPT Chat for Learning Purposes

Area	Positive Aspects	Negative Aspects
Formal education (school, high school, university)	- Availability 24/7 for helping with learning - Explains difficult concepts from subjects—student can do this in their own pace, ask questions - Helps in homework - Learning through conversation - Helps with writing essays, papers, and reports - Generates ideas for projects and presentations - Helps with quick review before tests or exams - Supports step-by-step problem solving	- Tendency to copy without understanding - Replaces critical thinking with ready-made answers - Difficult to verify accuracy of generated content - Risk of unintentional plagiarism - Weakens motivation to learn independently - Loss of patience and the ability to concentrate for long time - Difficulty undertaking tasks that require time and effort
Professional and digital skills development	- Speeds up learning new topics related to work - Assists in creating CVs, cover letters, and interview preparation - Simplifies technical terms and concepts - Supports learning programming or digital tools - Offers 24/7 access to expert-level explanations	- Lack of interaction with real mentors or practitioners - Knowledge may be too general or shallow - Risk of outdated or irrelevant advice - Overdependence on AI for learning - Oversimplifies complex subjects
Language learning and communication skills	- Enables conversation practice in many languages - Corrects grammar and language mistakes - Assists with translations and writing - Teaches idioms, cultural nuances, and formal vs. informal language - Helps practice real-life dialogues (emails, resumes) - Helps shy ones or people with language barriers to dare to talk	- No emotional depth in interaction - May provide incorrect or awkward language - Lacks cultural context in feedback - Style of writing can be too artificial or robotic - May reduce motivation to interact with real people
Personal growth and creativity	- Generates ideas for stories, art, creative writing - Helps plan daily tasks, goals, and reflections - Provides supportive "conversation," emotional support - Trains soft skills like communication and self-presentation - Facilitates self-reflection exercises	- May suppress original thinking - Risk of relying on AI for decisions - Surface-level support with limited emotional understanding - Reduces real-life inspiration and exploration - Can limit deeper emotional and cognitive development

Source: Author's own work.

participate in free workshops (Schirmer et al., 2023). Through these workshops, seniors learn to use smartphones, the internet, city e-services (doctor consultations, online payments, official apps), social networks, and data protection. Another way of increasing digital literacy is mobile digital learning points. For example, educational buses visiting various city districts, especially where there are poor transport connections or where there are older and migrant populations. In addition, social media campaigns, brochures in clinics and offices, and posters in offices can trigger learning and inform about courses on offer.

Another important feature of a Smart City in relation to people is inclusiveness. Inclusiveness can refer to various social groups and problems of exclusion. One of them is the older adults. It turns out that AI can be effectively used to support the care of seniors. It can support home care, social services, intelligent technical means (their service), and medical services. An author (Sun, 2021) presents a proposal for a "cloud service platform," the aim of which is to create a new model of community development suitable for aging populations. Such a solution includes an intelligent system of care services for the seniors, which will provide seamless connection between medical institutions, service providers, the elderly, and their families. This can offer daily assistance in simple activities, medical care, relaxation, and entertainment, as well as, importantly, emotional support. The proposed approach not only aims to improve the quality of life of the seniors but also to serve as an innovative platform for information management for the social care of the older adults. In addition, Smart Cities should be built technically and engineering-wise so that older adults have no problem using various technological means, contactless payments, or handling official matters. Smart Cities should also offer various support programs for older adults, educating them about digital literacy, and offering themed meetings and support groups. Consequently, Smart Cities should bridge the digital divide between the younger generation, who quickly adopt technological innovations, and older adults, who can often feel excluded.

Another important area where AI supports citizens is increasing the accessibility of services and public spaces for people with disabilities. AI can translate texts into sign language in real time, recognize speech and convert it to text, and assist blind people using computer vision systems (Thanukula et al., 2024). For example, mobile applications with AI enable visually impaired people to navigate the city by analyzing camera images and providing verbal instructions. AI-based systems can also assist people with intellectual disabilities in filling documents and searching websites by simplifying content and interfaces. Through such solutions, AI helps with providing social equality, supporting the independence and active participation of all city residents, despite their physical or cognitive abilities.

AI also supports the adaptation of physical spaces. Traffic management systems can take into account users in wheelchairs, older adults, or other people with limited mobility. For example, such systems can extend green light duration at crossings or suggest alternative, more accessible routes. Also, AI can be used for lighting in other places such as parks (Lis et al., 2024). AI can also analyze urban plans for space accessibility and suggest infrastructure changes—for example, adding ramps, eliminating thresholds, or adjusting the height of ticket machines. Solutions that support the consumption of content are increasingly common in museums, libraries, and theaters—for example, AI-powered voice guides, automatic subtitles for performances, or interactive stations accessible to people with various disabilities.

7.3 AI FOR PERSONALIZED PUBLIC SERVICES

The development of AI opens up new possibilities for improving and streamlining public services. Very often, residents have major problems with dealing with matters at the office, getting to a given specialist, or using digital solutions available in the city (e.g., parking fees). On the one hand, AI is supposed to help solve problems, and on the other hand, make public services more convenient and efficient for citizens. In this respect, we also talk about personalization, i.e., adjustment to the individual needs, lifestyle, health condition, or social situation of a specific person.

One of the most common examples is AI-supported health platforms. Work is underway to ensure that they offer residents individual preventive plans, reminders about tests, lifestyle analysis, and dietary recommendations based on data from wearable devices (Mukhopadhyay et al., 2022). This enables early diagnosis of chronic diseases and better health management of residents, especially the elderly and lonely. Such platforms can integrate data from various sources—such as smartwatches, mobile applications, or electronic health records—to create a personalized picture of a given user's health (Bizjak et al., 2017). This allows doctors to respond more quickly to worrying signals, while patients receive specific recommendations tailored to their lifestyle. Moreover, these systems can motivate preventive care and regular check-ups by reminding them at the right time and increasing health awareness of the society.

When it comes to public transport, systems that personalize travel routes are becoming more and more common. For example, there are applications that recommend the best travel options based on travel history, time preferences, and availability of means of transport (Chen et al., 2024). Additionally, AI can analyze road traffic and delays in public transport and then update routes. This can be used in real time to avoid traffic jams. People with disabilities can receive information about elevators, ramps, and low-floor buses—also—in real time (Nasser et al., 2024). For the seniors or those with limited mobility, these systems can suggest the most convenient stops and walking time to them (for example, taking into account terrain and weather conditions). AI can also adapt the interface of the transport application (it can increase the font size, enable voice mode, or simplify navigation). This can be done to better meet the needs of different users. Moreover, analysis of big amount of data can support cities in planning new transport lines and adjusting timetables to the actual needs of residents (Cao et al., 2019).

Everyday city services can also be personalized. An example is waste management (Bharti et al., 2022)—AI-supported applications can remind residents of the collection dates for specific waste fractions. They can also inform about changes in the schedule resulting from, for example, holidays. They can also be used to suggest (teach) best practices in waste segregation. Another area is energy management in social housing. Systems based on AI analyze individual electricity consumption profiles and can optimize the operation of devices in order to reduce costs without reducing the comfort of life. In crisis situations—e.g., a sharp increase in energy prices—such systems can automatically adjust heating or lighting settings, helping tenants save money in this area (Hou et al., 2019). Communication with the office is also personalized. Thanks to AI, residents can receive notifications about administrative matters—e.g., the need to extend a document or the deadline for paying tax. This facilitates contact with offices, increases the availability of public services, and reduces the risk of omitting important official matters. Systems can remind people of important document deadlines (e.g., ID card, vehicle registration), translate complicated official language into simpler ones, and suggest "paths to dealing with matters" tailored to the user's life situation. Elderly people can receive voice notifications, and disabled people can receive an interface tailored to their needs (e.g., enlarged text or sign language). What's more, AI can analyze citizens' most common questions and create automatic answers, reducing the waiting time to contact the office and relieving officials. In addition, they can receive information about local events. However, in order to avoid so-called spam, such information should be tailored to the user's preferences.

Another interesting and important application of AI is the personalization of urban participatory programs. AI systems can analyze the history of a citizen's involvement in social life (e.g., participation in consultations, submitted postulates, votes in the civic budget). Thanks to this, they can suggest topics or events that match their interests and values. A person interested in ecology will receive an invitation to consultations on urban greenery, and a socially active person—to local civic workshops. This makes it easier to maintain citizen involvement and build a conscious urban community.

Finally, it is worth mentioning the applications of AI in personalizing city information and public warnings. Currently, most often (i.e., in the traditional communication model), all residents receive the same information. This can be ineffective, but also irritating for citizens. Meanwhile, AI can be used to deliver only those messages that are actually important. For example, an air pollution warning can be sent to a person with respiratory diseases, or information about road renovations can reach a person who regularly travels a given route. Notifications can also be sent in the form of entertainment, e.g., based on musical preferences, a given person can be notified about a cultural event (e.g., a concert of their favorite band) in the area. Such communication, adapted to the needs and lifestyle of the recipient, increases its effectiveness and makes residents better aware of their surroundings and feel more cared for by the city.

In summary, AI is increasingly being used to personalize public services, making them more accessible, effective, and tailored to the individual needs of residents. AI should help make city services accessible to people, and not associated with problems—as is often the case today. AI-based solutions should support healthcare (e.g., through intelligent preventive platforms), transport (e.g., personalization of travel routes), management of everyday city services (e.g., energy, waste), or communication with offices. In the future, we can expect an increasingly wider implementation of such solutions, especially in the context of an aging society and growing expectations regarding the quality of public services. However, ensuring transparency and privacy protection will be crucial, as AI would be based on huge amounts of data, including sensitive data. Equal access to these services is also important, so that Smart Cities are truly inclusive and fair for all residents.

BIBLIOGRAPHY

Abdizoitovich, G. (2025). Educational problems in the information society. *Emergent: Journal of Educational Discoveries and Lifelong Learning (EJEDL)*, 6(4), 5.

Agrawal, S.S., and Panchal, S.B. (2023). The food app – fair and equal access to free food for anyone in need. *Proceedings of IDC 2023 – 22nd Annual ACM Interaction Design and Children Conference: Rediscovering Childhood*, 435–438. https://doi.org/10.1145/3585088.3595282

Bharti, S., Fatma, S., and Kumar, V. (2022). AI in waste management: The savage of environment. *Environmental Informatics: Challenges and Solutions*. Singapore: Springer Nature, pp. 97–123.

Bizjak, J., Gradišek, A., Stepančič, L., Gjoreski, H., and Gams, M. (2017). Intelligent assistant carer for active aging. *EURASIP Journal on Advances in Signal Processing*, 2017(1), 76. https://doi.org/10.1186/s13634-017-0511-y

Božić, V., and Poola, I. (2023). *Chat GPT and education*, p. 10.

Cao, Z., Ceder, A.A., and Zhang, S. (2019). Real-time schedule adjustments for autonomous public transport vehicles. *Transportation Research Part C: Emerging Technologies*, 109, 60–78.

Chen, A., Ge, X., Fu, Z., Xiao, Y., and Chen, J. (2024). TravelAgent: An AI assistant for personalized travel planning. *ArXiv Preprint ArXiv:2409.08069*. [preprint]

Feher, K. (2021). Expectation of smart mentality and citizen participation in technology-driven cities. *Smart Structures and Systems, An International Journal*, 27(3), 435–445.

Hou, X., Wang, J., Huang, T., Wang, T., and Wang, P. (2019). Smart home energy management optimization method considering energy storage and electric vehicle. *IEEE Access*, 7, 144010–144020.

Lis, A., Zienowicz, M., and Błachnio, A. (2024). Lighting features affecting the well-being of able-bodied people and people with physical disabilities in the park in the evening: An integrated and sustainable approach to lighting urban green areas. *Sustainability*, 16(20), 8871.

Mukhopadhyay, S.C., Suryadevara, N.K., and Nag, A. (2022). Wearable sensors for healthcare: Fabrication to application. *Sensors*, 22(14), 5137.

Nasser, N., Ali, A.Y., Karim, L., and Al-Helali, A.A. (2024). Enhancing mobility for the visually impaired with AI and Iot-enabled mobile applications. *ScienceOpen Preprints*. [preprint]

Obracht-Prondzyńska, H., Duda, E., Anacka, H., and Kowal, J. (2022). Greencoin as an AI-based solution shaping climate awareness. *International Journal of Environmental Research and Public Health*, 19(18), 11183.

Price, C.B., and Price-Mohr, R.M. (2018). An evaluation of primary school children coding using a text-based language (Java). *Computers in the Schools*, 35(4), 284–301.

Schirmer, M., Dalko, K., Stoevesandt, D., Paulicke, D., and Jahn, P. (2023). Educational concepts of digital competence development for older adults—a scoping review. *International Journal of Environmental Research and Public Health*, 20(13), 6269.

Streitz, N.A. (2021). Empowering citizen-environment interaction vs. importunate computer-dominated interaction: Let's reset the priorities! *Communications in Computer and Information Science*, 1351, 41–59. https://doi.org/10.1007/978-3-030-67108-2_3

Streitz, N.A., Riedmann-Streitz, C., and Quintal, L. (2022). From 'smart-only' island towards lighthouse of research and innovation. *International Conference on Human-Computer Interaction*, 105–126.

Sun, H. (2021). Research on the application of artificial intelligence technology and cloud computing in smart elderly care information platform. *2021 3rd International Conference on Artificial Intelligence and Advanced Manufacture*, 2451–2456.

Thanukula, K.L., Balanagu, R.K., Vasireddy, N., Gollamandala, G.C., and Pyla, A. (2024). Real-time communication for disabled people. *2024 International Conference on Emerging Systems and Intelligent Computing (ESIC)*, 597–602.

CHAPTER 8

Cybersecurity and Privacy in Smart Cities

8.1 IDENTIFYING AI-DRIVEN CYBERSECURITY THREATS

In times of rapid technological development, especially artificial intelligence (AI), cities are becoming increasingly dependent on ubiquitous connectivity, the Internet of Things, and cloud computing. This digital transformation brings many benefits, but it also exposes Smart Cities to new and increasingly sophisticated cybersecurity threats. AI, which can support protection against cyberattacks, can also be exploited by cybercriminals to carry out complex and difficult-to-detect activities.

One of the greatest threats is AI-powered attacks (Veprytska and Kharchenko, 2022; Waizel, 2024). Cybercriminals use machine learning to analyze victim systems, identify their weaknesses, and automatically create malware tailored to a specific target (Grispos, 2021). An example is AI-generated malware—code that learns the functioning of the environment in which it finds itself and adapts its form to avoid detection by traditional antivirus systems (Yedalla, 2024). Generative techniques also enable the rapid creation of new variants of viruses and Trojans. Many publications (Arafah et al., 2025; Chitimoju, 2023; Pa Pa et al., 2023; Yedalla, 2024) report that attacks using AI and autonomous hacking systems are increasingly common on the internet. This shift makes it harder for traditional defense methods to remain effective, as attackers leverage AI to launch cyberattacks and find vulnerabilities.

Another threat is deepfakes—images, audio, or video recordings created by AI models (e.g., GAN—Generative Adversarial Networks). This name is a combination of "deep," which refers to deep learning, and "fake" (Altuncu et al., 2024). They are used for identity theft (Gilbert and Gong, 2024; Lokhande et al., 2024), political manipulation (Battista, 2024), misinformation (Singh and Dhiman, 2023), and even money extortion (Volkova,

DOI: 10.1201/9781003587613-9

2025). An example of the last one can be impersonating a family member or a company supervisor—this is so-called deepfake phishing (Gal and Bulgurcu, 2024). In the context of cities, this can result in fake messages purportedly issued by local authorities.

The automation of phishing attacks is another consequence of cybercriminals' use of AI (Basit et al., 2021). Language models such as GPT can generate realistic and personalized emails that are more difficult to distinguish from authentic ones (Eze and Shamir, 2024). By analyzing public data (e.g., from social media), the content of messages can be made to specific individuals, increasing the effectiveness of attacks.

Attacks on IoT infrastructure are also on the rise. The subject of hacking can be the following: smart sensors (Tai et al., 2022), streetlights and traffic lights (Özarpa et al., 2021), surveillance cameras (Kronman, 2023), and even smart parking meters (Juliadotter, 2016). Due to their large number, diversity, and often weak security, these devices become easy targets for cybercriminals. AI can be used to scan large networks for vulnerabilities and coordinate massive distributed denial of service (DDoS) attacks from thousands of devices simultaneously, so-called botnets (Hoque et al., 2015).

There is also the risk of data poisoning (Schwarzschild et al., 2021). This threat can be understood as the manipulation of datasets used to train AI models. If such data is deliberately falsified (e.g., through faulty reports from residents to the traffic management system), it could lead to erroneous administrative decisions or erroneous recommendations made by the AI.

Also, internal attacks should not be forgotten, in which municipal employees use AI in an unauthorized manner or manipulate the results of predictive systems. Combined with the lack of transparency in algorithms, such actions can lead to violations of privacy, discrimination, or a loss of public trust. Table 8.1 presents examples of AI-driven cybersecurity threats and their potential consequences.

To effectively identify such threats, it is essential to invest in digital surveillance and anomaly detection tools that can monitor network traffic in real time and identify abnormal behavior patterns. Examples include solutions that leverage machine learning to predict and detect unusual activity early—for example, a sudden increase in network traffic or unusual system logins.

Furthermore, city authorities should regularly conduct security audits of AI-based systems. This allows for the identification of vulnerabilities before cybercriminals exploit them. It is also crucial for cities to collaborate with external experts and cybersecurity companies that offer red teaming, i.e., simulated attacks to verify the estimated security level. However, awareness of cyber threats is paramount, as without knowledge of AI and its associated threats, residents of Smart Cities will be unable to protect themselves from the threats discussed. Therefore, education of public administration employees and citizens is essential. Properly trained personnel will recognize the symptoms of an attack more quickly, and informed residents will not easily fall victim to manipulation or phishing scams. Cybersecurity in Smart Cities cannot be based solely on technology—it must also include organizational, educational, and preventive measures.

TABLE 8.1 Examples of AI-Driven Cybersecurity Threats and Their Potential Consequences

Type of Threat	Description	Potential Consequences
AI-generated malware	Malicious software dynamically created by AI that adapts to its environment	Bypasses traditional security systems; difficult to detect and eliminate
Deepfake audio/video	Fake content used to impersonate individuals or public officials	Disinformation, impersonation, public panic, loss of trust
AI-supported phishing	Highly personalized, realistic phishing emails generated by language models	Identity theft, account breaches, data leaks
Attacks on IoT infrastructure	Exploitation of vulnerabilities in connected city devices (e.g., cameras, sensors)	Network failures, traffic manipulation, public safety risks
AI-controlled botnets	Use of AI to coordinate mass-scale botnet attacks via infected devices	DDoS attacks, paralysis of public digital services
Data poisoning	Inserting manipulated or false data into AI training sets	Inaccurate algorithmic decisions, policy errors, skewed data analysis
Insider attacks	Misuse of AI systems by authorized personnel	Privacy breaches, manipulation of outcomes, reputational damage
Public opinion manipulation	AI-generated campaigns on social media to influence citizen behavior	Social polarization, misinformation, political interference
Password inference from media	AI reconstructs keyboard inputs from audio or video to extract passwords	Unauthorized account access, digital identity theft

Source: Author's own work.

8.2 STRATEGIES FOR SAFEGUARDING DATA AND ENSURING PRIVACY

With increasing digitization and the integration of AI into urban infrastructure, protecting citizens' personal data and privacy is becoming one of the key challenges for Smart Cities. In Smart Cities, data is collected almost continuously. This happens through surveillance cameras, mobile applications, IoT sensors, transportation systems, e-health platforms, and waste management. The environment is becoming increasingly digitalized. Consequently, data protection issues are becoming increasingly serious and difficult to solve. Such protection must be carefully planned and implemented comprehensively and at various levels. Its goal is to prevent abuse, information theft, unauthorized access, and unethical use of citizen information.

In Smart Cities, data has great value. It can be used for predictive analytics, service optimization, traffic management, and even administrative decision-making. On the other hand, when data are so much important, the issue of responsibility for data and saving it becomes more and more challenging. Leaks of medical, location, or behavioral data can have serious consequences. For example, they can lead to human rights violations (Predescu and Balan, 2023), discrimination, and even blackmail (Hussein et al., 2022). Therefore, in practice, taking care of data security must become one of the priorities in Smart Cities, where IoT and data monitoring have a continuous character. There are some strategies and policies that can be implemented. The examples are "privacy by design" and "security by default" policies (Bravo, 2021). They provide privacy protection from the beginning of the system design.

Regarding data protection strategies, the following are the most important:

- data encryption—data transmitted within Smart City infrastructure should be encrypted both in transit (in other words during transmission) and at rest (in other words on servers) (Arora, 2025). End-to-end encryption protects information from interception by third parties. For example, such a protocol could be used to protest data in communication between a smart air quality sensor and a city data center.

- access management and authorization—based on this approach only authorized individuals and systems should have access to data. Implementing multi-factor authentication (MFA), role management (specifically, role-based access control, RBAC), access login, and encrypted credentials help reduce the risk of internal abuse. Multi-factor authentication is a login method that requires more than just a password. For example, in addition to entering a password, users also need to enter a code from an SMS or confirm their login in a mobile application (Henricks and Kettani, 2019). This ensures that even if someone knows their password, they will not be able to log in without a second confirmation. Another policy—role management—ensures that each user in the system is assigned a specific role (e.g., employee, manager, administrator). Each role has different permissions, meaning they can see or do different things (Ferraiolo et al., 1995). This prevents, for example, an ordinary employee from accessing sensitive data that only a manager has access to. With the "access logging" approach, the system records who accessed which data and when (Preston, 2021). This allows for later verification of whether there was unauthorized access or whether someone abused their privileges. Within encrypted credentials, login data, such as passwords, are stored in the system in encrypted form (i.e., converted into a string of characters that cannot be read without a special key). This means that even if someone hacks into the system, they will not be able to read the passwords (Mustacoglu et al., 2020).
- data anonymization and pseudonymization—before analysis, personal data should be anonymized, which means that it cannot be attributed to a specific person. Pseudonymization is assigning a random identifier to a user so that they can be tracked without recognizing their identity. A system that analyzes residents' mobility without knowing their names can be an example of this policy.
- incident detection and response systems—real-time security monitoring enables rapid detection and response to breaches. Implementing Security Information and Event Management (SIEM) and Security Orchestration, Automation, and Response (SOAR) solutions allows for automated responses to threats (Măcăneață, 2024).
- regular audits and penetration tests—verifying system vulnerabilities and periodic security tests allow for the detection of weaknesses before cybercriminals do. Such measures should be considered good practice in managing Smart City infrastructure.
- compliance management (e.g., GDPR)—Smart Cities should operate in compliance with local and international data protection regulations. Established policies must be in place to inform citizens about data collection, processing methods, and their rights to modify or delete data.

It is important to remember that even the best data protection systems will be ineffective if end users (i.e., residents) are unaware of the threats. Therefore, it's worth implementing information and educational campaigns on digital privacy, strong passwords, safe application use, and phishing detection. Data protection begins at the individual level. Therefore, data protection is not only an IT issue but also a social one in Smart Cities.

While AI can be a source of threats, it can also serve as a security tool. AI-based systems can recognize attack patterns, automatically detect anomalies in data, and quickly respond to threats (Amarasinghe et al., 2019). For example, algorithms can detect unusual logins to city information systems or detect patterns of malicious network traffic. At the same time, AI can be used to protect privacy by applying privacy-preserving AI. They are models that learn from encrypted data or in distributed environments where user data is never transferred outside of their devices (Perino et al., 2022).

Protecting data and privacy in Smart Cities is a complex process. It requires the collaboration of technology, law, organizations, and education. In an age where every citizen's movement can be recorded, processed, and analyzed, responsibility for data cannot rest solely with end users. Cities must design their digital systems with privacy protection in mind from the outset. Otherwise, public trust in the Smart City concept could be seriously undermined, and the benefits of digitalization lost. The issue of monitoring such large amounts of data about people remains highly controversial and raises numerous issues related to data protection and citizen security, which will be difficult to resolve.

8.3 BUILDING RESILIENT CYBER INFRASTRUCTURES

As cities become increasingly intelligent, their digital infrastructure is simultaneously becoming more complex and vulnerable to threats. Millions of connected devices, real-time data flows, dependence on the cloud, and AI algorithms make cybersecurity a concept that is important to everyone, regardless of whether they use the internet or not. Because in today's world, even if people don't personally want to take advantage of the advances of digitalization, they will still encounter them, for example, through smart monitoring in their cities. With such advanced technology, it's important to emphasize that responding to incidents is no longer enough. Resilient systems are needed—those that can survive an attack, minimize its effects, and quickly return to normal operation. A resilient cyber infrastructure is the foundation of public trust and the security of public services in Smart Cities.

Resilience in cybersecurity is not just about avoiding attacks, but above all, the ability to survive them and quickly return to normal. According to Björck's definition cyber resilience is "*the ability to continuously deliver the intended outcome despite adverse cyber events*" (Björck et al., 2015). Today, no system is completely immune to threats—the question is not "if" an incident will occur, but "when." Therefore, a resilient infrastructure should skillfully detect threats, respond to them, limit damage, and recover. Such architecture helps avoid city paralysis in the event of, for example, a ransomware attack, server failure, or mass data leak.

The foundation of building a resilient infrastructure is layered security (Borkar, 2025). This means that security measures are not concentrated in a single location but distributed across multiple layers—from endpoint devices (IoT sensors), through local and cloud servers, to communication networks. Each layer should have its own security features: firewalls, intrusion detection systems (IDS), activity monitoring, access control, data encryption, network segmentation, etc. This means that even if one security measure fails, the next one provides the next line of defense.

It is also worth emphasizing that a resilient city does not passively wait for an attack, but should (indeed, must) constantly analyze data and signals coming from various sources. Systems using AI and machine learning can learn normal behavior patterns within the system and automatically detect anomalies—for example, unusual network traffic, logins at odd hours, or unauthorized data access (Gulhane and Sajana, 2021). AI algorithms create mathematical models based on sample or training data, used to predict unpredictable events or make decisions. Machine learning provides the ability to learn independently. In other words, a machine can learn from experiences, observations, and patterns analyzed from a dataset (Nozari and Sadeghi, 2021). The faster the detection, the less the damage. Threat prediction also enables earlier implementation of preventative measures.

Building a resilient infrastructure also requires human and organizational resources, including specialized Security Operations Centers (SOCs) (Vielberth et al., 2020). These centers analyze alerts, make decisions, and coordinate incident responses. The SOC team can immediately isolate the infected system, notify the appropriate services, and initiate data recovery. These centers should operate 24/7 and be equipped with advanced monitoring tools.

Every organization managing data in a Smart City must have a ready-made Incident Response Plan (IRP) and Business Continuity Plan (BCP) (Cochran, 2024). Such a plan should include: a list of emergency contacts, procedures for various types of threats, instructions for rapid data recovery from backups, and a test schedule for these procedures. Regular testing and updating plans are also important to avoid the system becoming inoperable in the event of a crisis.

Another issue to consider is segmentation. Not all parts of the city's infrastructure should be directly connected. Network segmentation (i.e., dividing it into smaller zones) can limit the spread of threats (Kallatsa, 2024). For example, even if a hacker gains access to a streetlight system, they shouldn't have access to residents' health data or central energy management systems. This is supported by appropriate permission management—each unit or system has access only to the data and functions it needs. Table 8.2 presents a summary of the above-mentioned key elements of a resilient digital infrastructure in a Smart City.

Building resilient infrastructure isn't the task of a single office or company. It requires the collaboration of many stakeholders—city authorities, digital service providers, public institutions, and residents. Implementing uniform cybersecurity standards is crucial. Such standards help create a coherent operational framework and improve the level of protection across the entire city.

In the future, the level of automation of resilient cybersecurity systems will be higher. They will use not only AI but also technologies such as quantum encryption (Xu et al., 2021), which allows for virtually unbreakable data transmission. Self-learning systems will also play a more important role and will be focused on adapting security policies to the threats in the given environment. This will help cities to operate efficiently, even in growing digital risk.

TABLE 8.2 Summary of the Key Elements of a Resilient Digital Infrastructure in a Smart City

Area	Description of Functionality
Layered Security	Implementing protection at every level of infrastructure—from endpoint devices to the cloud
Early Threat Detection	Anomaly detection, AI-powered monitoring, and Intrusion Detection Systems for fast alerts
Security Operations Center (SOC)	24/7 monitoring, real-time alert analysis, and rapid incident response coordination
Emergency and Recovery Plans (BCP/DRP)	Procedures for data recovery, crisis management, and ensuring business continuity
Network Segmentation and Access Control	Dividing networks into zones, controlling access rights, and minimizing data exposure

Source: Author's own work.

In summary, resilient cybersecurity infrastructure is the foundation of a Smart City. Without it, it can be difficult to provide security of residents' data and the continuity of key services. Therefore, investments in this area should not be viewed as a cost, but rather as a strategic foundation for a modern, digital society.

BIBLIOGRAPHY

Altuncu, E., Franqueira, V.N.L., and Li, S. (2024). Deepfake: Definitions, performance metrics and standards, datasets, and a meta-review. *Frontiers in Big Data*, 7, 1400024.

Amarasinghe, A., Wijesinghe, W., Nirmana, D.L.A., Jayakody, A., and Priyankara, A.M.S. (2019). AI based cyber threats and vulnerability detection, prevention and prediction system. *2019 International Conference on Advancements in Computing (ICAC)*, 363–368.

Arafah, M., Al-Banna, A.-K., and Aladawi, A. (2025). Detection and analysis of ai-generated malicious content. In *Examining Cybersecurity Risks Produced by Generative AI* (pp. 235–276). IGI Global Scientific Publishing.

Arora, A. (2025). Analyzing best practices and strategies for encrypting data at rest (stored) and data in transit (transmitted) in cloud environments. Available at SSRN 5268190. [preprint]

Basit, A., Zafar, M., Liu, X., Javed, A.R., Jalil, Z., and Kifayat, K. (2021). A comprehensive survey of AI-enabled phishing attacks detection techniques. *Telecommunication Systems*, 76(1), 139–154.

Battista, D. (2024). Political communication in the age of artificial intelligence: An overview of deepfakes and their implications. *Society Register*, 8(2), 7–24.

Björck, F., Henkel, M., Stirna, J., and Zdravkovic, J. (2015). Cyber resilience–fundamentals for a definition. In *New Contributions in Information Systems and Technologies*: Volume 1 (pp. 311–316). Springer.

Borkar, M.J.P.K. (2025). The Cloud Defense: Building Resilient Security Layers.

Bravo, F. (2021). Data management tools and privacy by design and by default. In *Privacy and Data Protection in Software Services* (pp. 85–95). Springer.

Chitimoju, S. (2023). The risks of ai-generated cyber threats: How LMs can be weaponized for attacks. *International Journal of Digital Innovation*, 4(1), 1–8.

Cochran, K.A. (2024). Incident response, business continuity, and disaster recovery. In *Cybersecurity Essentials: Practical Tools for Today's Digital Defenders* (pp. 385–430). Springer.

Eze, C.S., and Shamir, L. (2024). Analysis and prevention of AI-based phishing email attacks. *Electronics*, 13(10), 1839.

Ferraiolo, D., Cugini, J., and Kuhn, D.R. (1995). Role-based access control (RBAC): Features and motivations. *Proceedings of 11th Annual Computer Security Application Conference*, 241–248.

Gal, S., and Bulgurcu, B. (2024). Exploring Factors Influencing Internet Users' Susceptibility to Deepfake Phishing, AMCIS Proceedings, Salt Lake City, 21.

Gilbert, A., and Gong, Z. (2024). Digital identity theft using deepfakes. In *Information Technology Security and Risk Management* (pp. 307–314). CRC Press.

Grispos, G. (2021). Criminals: Cybercriminals. In *Encyclopedia of Security and Emergency Management* (pp. 84–89). Springer.

Gulhane, M., and Sajana, T. (2021). Human behavior prediction and analysis using machine learning-A review. *Turkish Journal of Computer and Mathematics Education*, 12(5), 870–876.

Henricks, A., and Kettani, H. (2019). On data protection using multi-factor authentication. *Proceedings of the 2019 International Conference on Information System and System Management*, 1–4.

Hoque, N., Bhattacharyya, D.K., and Kalita, J.K. (2015). Botnet in DDoS attacks: Trends and challenges. *IEEE Communications Surveys & Tutorials*, 17(4), 2242–2270.

Hussein, O.A., Manap, N.A., Rahman, M.R.A., and Wahhab, B.M.A. (2022). Cyber blackmail on social media and its authenticity through criminal evidence. *NeuroQuantology*, 20(6), 121–132.

Juliadotter, N.V. (2016). Hacking smart parking meters. *2016 International Conference on Internet of Things and Applications (IOTA)*, 191–196.

Kallatsa, M. (2024). *Strategies for Network Segmentation: A Systematic Literature Review*. Jyväskylä, Finland: University of Jyväskylä, pp. 1–76.

Kronman, L.M.J. (2023). Hacking surveillance cameras, tricking AI and disputing biases: Artistic critiques of machine vision. *Open Library of Humanities*, 9(2), 1–35.

Lokhande, M., Raut, P., Gawali, K., Ahirrao, M., and Bhande, A. (2024). Artificial intelligence for detecting cyber attacks in deepfake & identity theft. *2024 8th International Conference on Computing, Communication, Control and Automation (ICCUBEA)*, 1–6.

Măcăneață, C. (2024). Overview of security information and event management systems. *Informatica Economica*, 28(1), 15–24.

Mustacoglu, A.F., Catak, F.O., and Fox, G.C. (2020). Password-based encryption approach for securing sensitive data. *Security and Privacy*, 3(5), e121.

Nozari, H., and Sadeghi, M.E. (2021). Artificial intelligence and machine learning for real-world problems (a survey). *International Journal of Innovation in Engineering*, 1(3), 38–47.

Özarpa, C., Avcı, İ., Kınacı, B.F., Arapoğlu, S., and Kara, S.A. (2021). Cyber attacks on scada based traffic light control systems in the smart cities. *The International Archives of the Photogrammetry, Remote Sensing and Spatial Information Sciences*, 46, 411–415.

Pa Pa, Y.M., Tanizaki, S., Kou, T., Van Eeten, M., Yoshioka, K., and Matsumoto, T. (2023). An attacker's dream? Exploring the capabilities of ChatGPT for developing malware. *Proceedings of the 16th Cyber Security Experimentation and Test Workshop*, 10–18.

Perino, D., Katevas, K., Lutu, A., Marin, E., and Kourtellis, N. (2022). Privacy-preserving AI for future networks. *Communications of the ACM*, 65(4), 52–53.

Predescu, P.-A., and Balan, D. (2023). The implications and effects of data leaks. *International Conference on Cybersecurity and Cybercrime*, 10, 170–177.

Preston, W.C. (2021). *Modern Data Protection*. O'Reilly Media, Inc.

Schwarzschild, A., Goldblum, M., Gupta, A., Dickerson, J.P., and Goldstein, T. (2021). Just how toxic is data poisoning? A unified benchmark for backdoor and data poisoning attacks. *International Conference on Machine Learning*, 9389–9398.

Singh, P., and Dhiman, D.B. (2023). Exploding AI-generated deepfakes and misinformation: A threat to global concern in the 21st century. Available at SSRN 4651093. [preprint]

Tai, R., Lin, L., and Su, R. (2022). Identification of system vulnerability under a smart sensor attack via attack model reduction. *IEEE Control Systems Letters*, 6, 2948–2953.

Veprytska, O., and Kharchenko, V. (2022). AI powered attacks against AI powered protection: Classification, scenarios and risk analysis. *2022 12th International Conference on Dependable Systems, Services and Technologies (DESSERT)*, 1–7.

Vielberth, M., Böhm, F., Fichtinger, I., and Pernul, G. (2020). Security operations center: A systematic study and open challenges. *IEEE Access*, 8, 227756–227779.

Volkova, S. (2025). The dark side of deepfakes: Fraud and cybercrime. In *Deepfakes and Their Impact on Business* (pp. 221–242). IGI Global Scientific Publishing.

Waizel, G. (2024). Bridging the AI divide: The evolving arms race between AI-driven cyber attacks and AI-powered cybersecurity defenses. *International Conference on Machine Intelligence & Security for Smart Cities (TRUST) Proceedings*, 1, 141–156.

Xu, D., Yu, K., and Ritcey, J.A. (2021). Cross-layer device authentication with quantum encryption for 5G enabled IIoT in Industry 4.0. *IEEE Transactions on Industrial Informatics*, 18(9), 6368–6378.

Yedalla, J. (2024). AI-generated cyber threats the rise of autonomous hacking systems. *International Journal of Advanced Research in Computer and Communication Engineering (IJARCCE)*, 13(12), 7.

CHAPTER 9

Case Studies of AI Usage in Smart Cities in Central Europe

9.1 SUCCESSFUL AI IMPLEMENTATIONS IN CENTRAL EUROPEAN CITIES

This chapter reviews and compares the status of Artificial Intelligence (AI) implementation in three important Central European cities: Warsaw, the capital of Poland; Prague, the capital of the Czech Republic; and Budapest, the capital of Hungary. These selected cities are of great cultural, economic, and technological importance in their region and play a pivoting role in their country's AI strategy and developments. It looks at the AI phenomenon along dimensions of government policy about AI, research and innovation related to AI, industry, and Smart City initiatives with regard to AI adoption. Warsaw, when it comes to supporting a plethora of new technologies, goes way back. It has just been confirmed by the report "State of Warsaw AI," which says that this capital joined the line of leading AI development centers both in number and value of investment in groundbreaking solutions. Presently, about 6,500 AI specialists work in this city, while their number is growing at a rate of about 54% annually. As many as 92% of employees have higher education, mainly technical, and 86% speak English. Among the several reasons investors are willing to develop high-tech firms, qualified work occupies a top place. Eighty-three of the employees in the capital's AI sector are quite convinced that a well-paying job could be found only in Warsaw in this industry. Major spheres of employment include IT at 35%, trade and e-commerce at 13% each, along with banking and finance at 13% each (Standards for Responsible Use, 2023).

Also, solutioning by AI was introduced in the solving of urban problems with the aim of making the administration more functional, thereby helping to enhance lifestyles.

DOI: 10.1201/9781003587613-10

Warsaw proactively outlined the ground rules that would allow enabling AI technologies during an ethical implementation process. It detects the possible risks that come with AI—most recently so-called GenAI—by adopting "Standards for Responsible Use of Generative AI in the City of Warsaw"; it underlines protection for data privacy, confirms GDPR, and makes the AI system work on its algorithms to become understandable and auditable. It also addresses the treatment of subjects for equality, non-discrimination, lack of algorithmic bias in treating subjects, so that all citizens are included (Standards for Responsible Use, 2023). Equally, the decision-making empowered by the AI systems must be possible and lines of responsibility demarcated clearly. The ethical framework was worked out in cooperation with the AI experts, academic researchers, and industry stakeholders and underlines Warsaw's commitment to the responsible governance of AI (Sztuczna inteligencja, 2023).

Such events like "Technological Reviews," where investors in the city of Warsaw showcase AI-driven technologies, do take place rather often. In urban monitoring that is powered by AI, aerial images are analyzed for real-time tracking of infrastructure, greenery, or even urban development in order to plan the city or to plan the conservation of the environment. Intelligent traffic management systems use AI algorithms in an effort to improve flow, reduce congestion, and efficiently use public transportation through the understanding and trying to avoid bottlenecks assisted by real-time data. Innovations in waste management involve intelligent bins fitted with sensors and AI algorithms, which manage waste collection scheduling optimally for further facilitation of recycling, enabling both economic and ecological benefits. Automation of document processing, by reading, classifying, and structuring data, will decrease the administrative burden further, contributing to effectiveness in the delivery of public services. AI-based surveillance systems will cater to the safety and security of public life, since it is possible to trace the anomalies and threats in view of creating a safer urban space without breaching privacy concerns. Such applications only point to Warsaw's strategic spotlight in terms of sustainability, efficiency, and safety (Sztuczna inteligencja, 2024; Sate of Warsaw, 2024).

This has resonated in the "State of Warsaw AI" report, prepared by the Digital Poland Foundation: pointing out the grave economic impact brought about by companies connected with AI, whose contribution to the capital's economy is extremely high, employing over 6,500 specialists, in addition to international investments that have been attracted to the city. With an average growth rate of 54% YoY, the Warsaw AI market demonstrates the capital as a technological leader for Central and Eastern Europe. A number of research centers here work together with universities for the development of AI-related inventions. Venture capital sweetens investment opportunities both for startups and established companies by investing government funds into startups operating in this field and accelerates its adoption and development. This report underlines further that the focus of Warsaw on research and development is away from commercialization (Standards for Responsible Use, 2023). The Warsaw AI Community is a grassroots organization to develop AI and knowledge of AI for the enrichment of Warsaw's AI ecosystem. This kind of community could arrange meetups that can make operating people in AI interact with researchers and policymakers. Events of networking might also be included. These workshops and training programs go forth to impart practical skills to participants or insights into the AI technologies. Competitions and hackathons spur creativity and problem-solving with a view to developing innovative solutions that can meet real-world challenges. Collaboration with startups and corporations bridges the gap between the emerging company and established business, therefore encouraging partnerships or joint ventures. With these initiatives, Warsaw.AI strengthens the local AI community and positions Warsaw as a vibrant hub of AI innovation (Warsaw AI, 2024).

Examples of implementation of AI in Warsaw, divided into particular Smart City areas, are given in Table 9.1.

The capital city of Prague is one of the most important centers in the elaboration and application of AI in the Czech Republic. Logically, it should seriously invest in research, education, and support for businesses—a strategic ecosystem contributing to the growth of AI with ethical consideration and international cooperation.

TABLE 9.1 Examples of AI Implementation in Warsaw

Smart City Area	Initiative	Description
Smart governance	E-Office (E-Urząd)	It will enhance the administrative efficiency of the system, reduce paperwork, and give a faster pace to the residents in whose hands the services applied for online and the taxes to be paid also get processed through digital platforms wherein document processing and query management have been enhanced by the incorporation of Artificial Intelligence.
Smart economy	Startup hub platform	It would have an environment for innovative start-up services that will satisfy various entrepreneurial needs, from mentorship and funding opportunities to creating AI-powered tools such as market insights, business optimizations, and predictive modeling. It really will close the gap in economic development through digital transformation and technological innovation as well.
Smart mobility	Intelligent Traffic Management System (ITS)	It will install AI-powered systems for real-time traffic flow monitoring and management, along with traffic light signals optimized for reducing congestion. It also offers integration with public transportation timetables and eco-friendly modes of transportation, such as cycling and electric vehicles, toward advanced urban mobility and sustainability.
Smart environment	Smart energy and water meters	Advanced deployment of intelligent energy and water meters monitors real-time resource consumption, which is helpful for finding out inefficiency and reducing waste in order to be environmentally friendly through energy conservation and reduced carbon emission.
Smart living	Health 2.0 platform	It will allow the development of digital health tools in preventive healthcare, telemedicine services, and remote health monitoring using AI technologies. Since the early diagnosis that this platform would provide means a resident could have more personalized treatment recommendations, access to healthcare for a quality life would improve.
Smart people	Digital and AI education programs	It would find realization in training programs, online courses, and workshops on raising digital literacy and AI-related competencies among the residents. These would foster not only inclusiveness but also work readiness in building citizens who can easily fit into the technology-driven economy and society.

Source: Author's own work on basis: (Standards for Responsible Use, 2023; Method, 2023; Construction, 2023; National Debut, 2024; State of Warsaw, 2024).

For example, PRG.ai, although based in Prague, is becoming a non-profit organization that aims to make the city one of the largest AI centers in Europe. This will bridge the gap from academic research to commercial application involving university-business-government collaboration. The incubation, acceleration, and mentoring in opportunities for AI startups are enabled to create events, thus making AI talent more aware of how this innovative technology is about to change industries (Transforming, 2023; Weber, 2024).

Other recent developments include the launching of the IBM Center of Excellence for Generative AI in Prague this 2023, enabling companies to make sure AI is applied in such a way that security, privacy, and fairness are central to business operations. This would help address the growing demand for responsible practices in the development and implementation of rapidly expanding AI technologies, ensuring that their use aligns with established ethical standards. It is also advising the use of AI to enhance productivity by optimization of resources spent by businesses. Besides Prague, an Artificial Intelligence Centre (AIC), attached to the Czech Technical University, will be relevant in the developments and research of this topic. Special interest will be developed in robotics, machine learning, cyber-security, and independent systems (Kolisnychenko, 2023; Empowering, 2024). Through industrial partners, AIC develops not only next-generation solutions but educates a new wave of professionals in the area of AI. In all, the programs at AIC would be a combination of academics, professional development, and research, offering students the set of all skills that would be in demand for the AI-driven economy. Education thus plays the core part of Prague's AI strategy. This is put together with high labor force demands for such skills: the universities have opened special programs in AI and data science, starting from undergraduate to graduate and PhD studies, but also courses for re-qualifying the work force that feels threatened by automation in some areas. Besides, workshops, hackathons, and other activities supporting AI add to the further dissemination of knowledge and innovation that will be required to respond to the challenges and opportunities created by AI technologies (Krasavina, 2023).

Furthermore, the effort by Prague extends to international cooperation in the establishment of the ethical framework concerning AI. Similarly active in European-level initiatives related to AI, the country pursues regional-level policies in equity, transparency, and protection of personal data. Legal and regulatory frameworks are being devised so that the technologies of AI respect human rights, observe standards of ethics, and contribute to building public trust in the systems of AI (National Artificial Intelligence, 2024).

This will give more insight into the AI that Budapest, as one of the leading cities, has to embed into various urban systems because of efficiency, sustainability, and innovation. Its application can also be seen in transport, healthcare, education, and town planning. The great effort given by the city of Budapest is in using AI in the concern for road safety. Equipped with the power of AI, full-fledged analysis was performed for mammoth data volume with regard to flow, accident history, and weather. With this system, accident-prone spots will be predicted along with the necessary propositions for preventive measures through predictive analytics. A good example is that, in October 2024, the authorities changed the speed limit on some roads along Attila's main street from 50 to

40 km/h based on an AI assessment. As a result, annual accident rates dropped by more than 50%. Parallel to the control of speed, Budapest uses smart traffic lights fitted with AI sensors. These in turn change dynamically according to congestion level, allowing smoother flow and less time spent waiting at the junctions. In this way, the system reduces idle time for cars and increases efficiency through a reduction in vehicle emissions. AI works in predictive maintenance, whereby the road infrastructure is always monitored against problems identified before they actually become disruptions. Such technologies allow for proactive fixing of the problems in order to make transportation safer and more reliable (Pesti, 2024).

Public transportation in Budapest, run by Budapesti Közlekedési Zrt., or BKV Zrt., is being transformed with the help of AI technologies. Changes are made so that public transportation becomes reliable and efficient, easy for passengers to use. It does this through intelligent real-time capture of information related to the location of vehicles, passenger flow, and demand patterns. Extracted information will in turn enable operators to undertake route planning and construction of timetables that guarantee smooth and punctual running of bus, tram, and metro services (Revolutionizing, 2024).

AI algorithms also forecast breakdowns of both vehicles and infrastructure, creating time for maintenance that helps reduce disruption to services. Further, installation of AI tools in mobile applications provides real-time arrival times, suggestions concerning alternative routes, and possible delays. Such innovations make commuting more predictable, decreasing uncertainty and, therefore, making the use of public transportation more attractive to both residents and visitors (Continental, 2018).

Recent investments by big players positioned Budapest as a development center of autonomous driving. German giant Continental opened a Competence Center for Deep Learning in Budapest with the task of developing AI for independent cars. More than 100 experts in machine learning and AI develop solutions with the seal of safety and efficiency first in this center (AI Hungary, 2020).

From object detection to lane recognition, even down to emergency braking, the AI-powered systems engineered in Budapest can do it. Giant-sized datasets run with sophisticated algorithms and simulate on-road scenarios to empower the cars for split-second decisions without human interference. In so doing, Budapest will overtake its current position among the emerging leaders of AI innovation and international investment and talent in the autonomous driving sector.

The first signs of the new faces of medical diagnosis started to appear in Budapest's health sector. Since March 2023, one of the most advanced research centers specializing in cancer diagnostics, the Pathology Innovation Incubator, has been in operation. The project required the collaboration of two companies with extensive expertise in the field—3DHISTECH Kft. and Epredia—for the installation of a state-of-the-art medical imaging facility. Therefore, it would involve the development of AI-driven systems, which, in turn, are set to support and enable pathologists to analyze medical images in a speedy and highly accurate manner. For example, just one AI can scan just one tissue sample for abnormalities in a few minutes. AI increases professionals' efficiency by automating most of the mundane tasks and reducing human error to deliver results that are far more reliable. The incubator wants to cut down diagnosis time as much as 50% in search of enabling early detection and treatment of diseases such as cancer (Budapest, 2023).

While this is Budapest arming for an AI-driven future, preparation of its workforce also means offering academic programs in AI itself: Free time To the Budapest Metropolitan University. This is going to be quite an exciting time because it is going to introduce this special program commencing in September 2024 for teaching the students the use of AI for practical purposes in business, finance, and technology. It will be taught through practice and fundamental theory with a view to giving such students the necessary preparation to be immediately competitive in today's dynamic labor market (Brader, 2024).

In this vein, AI-related courses are invited to focus even more on project-based learning with industrial partners to develop solutions that start from machine learning to data analysis and programming. Besides that, such a curriculum would not only prepare them to work in companies but also enable them to become entrepreneurs by creating AI-based startups. This will build a new generation of professionals in Budapest who can further contribute to the growth of this city technologically (Brader, 2024).

The Hungarian national AI strategy, presented early in 2020, was quite ambitious regarding the introduction of AI into the country's economy and society. It clearly speaks to increasing the GDP of the country by 15%, enhancing productivity by 26%, and retraining one million workers for AI-enhanced jobs until 2030. Budapest is the capital and, as such, it plays a major role in reaching these goals (Brader, 2024).

It was followed by the establishment of research facilities that would further contribute to innovation and development, including the National AI Laboratory. Besides this, academia-business-government agency collaboration is provided for quicker adoptions of AI. Financial incentives and grants also reach startups and firms developing products and services based on AI. With these moves, Hungary hopes it will be among the regional leaders in matters concerning AI technologies and Digital Transformation.

Table 9.2 presents a comparative overview of AI implementation in three cities under review: Warsaw, Prague, and Budapest.

9.2 CHALLENGES FACED BY CENTRAL EUROPEAN SMART CITIES

AI implementation in the Smart Cities of Central Europe, among which are Warsaw, Prague, and Budapest, is going to face a number of obstacles. Among the key hurdles, data privacy and security are considered major ones: most AI applications require huge volumes of data, including personal data; therefore, strict regulations, such as those promulgated by the European Union under the General Data Protection Regulation, must be maintained (Ojha and Nikhil, 2024). Another key challenge common among all three cities is to keep intact the privacy of citizens with the deployment of AI systems in transportation, health, and urban management. They are working their way out by investing in strong policies regarding data protection, upgrading their cybersecurity infrastructure, and conducting campaigns to make the general public aware of how the processing of their data will take place. At the same time, cooperation with the relevant EU bodies was an additional guide on how to correspond to the last standards of privacy (Jonek-Kowalska and Wolniak, 2023, 2025). The other challenge is the technological infrastructure of

TABLE 9.2 AI implementation—Warsaw, Prague and Budapest

Aspect	Warsaw (Poland)	Prague (Czech Republic)	Budapest (Hungary)
Government support and policy	Nowadays, Poland has begun to increase investment in developing AI through various national strategies, such as the AI Strategy for Poland 2020–2030. In its implementation, an ecosystem for AI, support for research, and foreign investments are in place. It is far slower in implementation compared to Western Europe.	The Government of the Czech Republic has a National Strategy on AI, putting much stress on research and innovation. It looks into close collaboration between the public and private sectors. There is an institutionalized framework for integrating AI in a host of governmental services and planning for cities. Being the capital, Prague stands up front in policy implementation.	In 2018, the Hungarian government adopted a National Artificial Intelligence Strategy aimed at developing AI further in the country. Certain concrete steps have indeed been taken in this direction, with the government encouraging AI research and applications in health and Smart Cities. Hungary has assumed a much more centralized strategy compared to neighbors that are far ahead in their implementation of AI.
AI research and education	Equipped with top-class research institutions like the University of Warsaw, the Polish Academy of Sciences, and Warsaw University of Technology, it has AI departments carrying out advanced research. Currently, the city is working on increasing the number of AI professionals, together with developing more collaborations with international institutions.	The research environment is already here in Prague; good academic institutions like the Czech Technical University and Charles University are already present. Moreover, the Czech Republic is in a good position in AI research, above all in robotics and data processing; it includes close connections with European research networks.	Budapest involves world-class universities, like Eötvös Loránd University and Budapest University of Technology and Economics, in high-level AI research. Traditionally strong in mathematical sciences, Hungary feeds into the innovation of AI mainly in algorithms and machine learning.
AI startups and innovation	The Warsaw startup ecosystem is fast developing, with several promising AI ventures in Fintech, Healthcare, and Enterprise AI. Poland is considered to be a developing hub in Central Europe for AI, though still in its nascent stages.	Prague is swiftly turning into a strong developing center for AI startups, pushed forward by innovative companies. Also, it enjoys a very good business climate and thus tends to attract international AI talent. The AI startup scene is well below the size of Warsaw and Budapest.	Budapest is developing an ecosystem that ranges from AI solutions for autonomous driving through to data analytics and automation. Smaller than Warsaw, Budapest nonetheless draws very strong local support and is an emergent focal point of AI activity in Central Europe.
Industry adoption	AI finds its way into manufacturing, finance, and logistics in Poland. Leading companies have already implemented the technology; for example, PKO Bank Polski started using it to cut operational work. Besides that, the government is encouraging more AI integration into public services and administration, although large-scale deployment in traditional industries remains very minimal.	The Czech Republic is a leader in the region in everything concerning the adoption of AI, particularly in the automobile, energy, and robotics industries. Prague has relevant AI involvement from big multinational companies like Škoda Auto and IBM about AI for production efficiency, Smart Cities, and numerous others. AI adoption within Prague is relatively advanced due to the involvement of multinational companies.	AI has already found applications in automotive, manufacturing, and healthcare industries in Hungary. For example, AImotive develops software solutions for autonomous driving, and healthcare AI solutions also start to appear. Compared with Prague, however, Hungary faces more challenges in terms of large-scale industrial AI adoption.

(*Continued*)

TABLE 9.2 (Continued)

Aspect	Warsaw (Poland)	Prague (Czech Republic)	Budapest (Hungary)
Smart City initiatives	Most of the projects in relation to Smart City initiatives are run by AI. Especially, much attention is given to the role of Artificial Intelligence in the Warsaw Smart City Strategy, within applications regarding urban management for the optimization of traffic flow and the rise of citizen safety. Nowadays, it is being integrated into the aspects of urban mobility, planning, and infrastructure.	Among Smart City initiatives, Prague is definitely one of the most aggressive, integrating AI into urban mobility, energy management, and public safety. A number of various AI applications have been installed in this city for traffic management, air quality, and waste management. The Czech capital is among the most developed Smart City projects in the region.	Most of the Smart City initiatives in Budapest are at their early stage, but there are some noticeable AI-powered projects concerning smart public transport optimization, lighting, and waste management systems. The city is progressing somewhat slower in comparison with Warsaw or Prague but will go further into the process of including AI in the administration gradually.
Investment and funding	The country hosts growing foreign and domestic investments in AI, especially in fintech and logistics verticals. While venture capital funding is growing in Warsaw, more significant foreign investment is still lagging compared to Prague.	Most of the investment has gone into robotics, automotive AI, and big data. Prague sees a fair amount of financing from both the Czech government and EU funds, but the ecosystem is considerably smaller than the big Western European capitals.	Most of the interest in AI investment has been directed at Budapest, with an especial focus on autonomous cars and healthcare AI. With strong academic foundations in place, Budapest has thus far failed to make key gains in international investments that compare to its neighbors, with funding still at a very developing stage.
Public awareness and society	Awareness of AI is growing in Warsaw: education, mass media, and technology communities bring more and more attention to the potentials of AI. Still, the sphere of public debate concerning the ethical consequences of AI needs development.	In Prague, AI is much better known and accepted than in academic and business circles, but it is simultaneously raising some concern about data privacy and ethics in using AI, already starting to be discussed in public debates.	Awareness of AI issues is gradually improving in Budapest, especially within the tech community and in business circles. Public engagement is growing, though the societal debate about the implications for employment or privacy has yet to be more intensive.

Source: Author's own work on basis: (Standards for Responsible Use, 2023; Method, 2023; Construction, 2023; National Debut, 2024; State of Warsaw, 2024; Krasavina, 2023; Stofan, 2024; Transforming, 2023; Pesti, 2024; Hungary's Artificial Intelligence, 2020; Budapest, 2023; Smart Budapest, 2024; Experience the Future of AI, 2024).

AI systems, high-speed internet, and other sophisticated means of communication. In Warsaw, Prague, and Budapest, there are some bottlenecks in the digital infrastructure that may press a brake on the wide dissemination of AI solutions. These cities address this issue through active modernization of their infrastructures, establishing fiber optics and 5G networks. Public-private partnership is also very important in developing and upgrading smart infrastructure; the gradual deployment of the AI systems ensures that cities can test and improve technology readiness over time (Ali et al., 2025).

The second challenge toward the implementation of AI is a lack of professionals in the field (Wolniak and Stecuła, 2024). The demand for data scientists, machine learning engineers, and AI specialists far exceeds supply in Central Europe. The cities create an alliance with universities and research institutions to promote education and training in AI. Furthermore, offering incentives for international talent and encouraging AI startups to set up businesses in their cities attracts the needed skills.

Other challenges are of a financial nature: AI projects have very high costs. Implementations on a larger scale than small-scale pilot projects are difficult to finance in cities confronted with limited public budgets. The cities are nonetheless considering solutions—that is, searching for funding from the EU for innovative projects or seeking participation on the part of the private sector in jointly co-funding AI activities. These cities indirectly invest in transportation and planning; they invest in supporting AI development and reinforcing any resources toward its further development (Singh and Nayyar, 2024).

AI systems also face the challenge of the integration process, involving pre-existing, at times quite outdated, infrastructures. Most of the Smart City solutions need interfaces with existing legacy systems that can be quite costly and a heavy burden on technology investment. This is what Warsaw, Prague, and Budapest try to respond to by following a step-by-step procedure of gradual phase integration. Full deployment based on pilot testing enables modular AI systems to adapt to older infrastructures. This is a flexible solution for mitigating the risk of disruption, with a guarantee of long-term scalability (Balashova and Reshetnikova, 2024). The other challenges facing these cities relate to the resistance from the public to AI. Most citizens are skeptical of AI, since they think it would intrude into their private life, make them lose their jobs, or change the social way of life. The cities counter this resistance by public consultations, showcasing successful AI implementations in other cities, and involving citizens in the decision-making process. The likely impact of these approaches is to build trust through demonstration of the tangible outcomes that AI can deliver (Estebsari and Werna, 2022).

Other challenges include regulatory and legal barriers. One of the reasons for uncertainty regarding the issues of liability, ethics, and decision-making processes in AI systems is a non-existent or inconsistent regulatory framework. Warsaw, Prague, and Budapest are working with EU institutions to align local regulations with the more general European guidelines. Moreover, cities are already setting up AI task forces and crafting ethical guidelines that can provide a better framework for governing the use of AI (Obaideen et al., 2024).

Not less important are issues of scalability and adaptability. While many AI systems function perfectly on a small scale, scaling them up across an entire city—or adapting them to the rapidly changing conditions—is quite a challenge. It was with this issue in mind that these cities came to be determined to create scalable and adaptable AI solutions. Pilot programs can test scalability before scaling up, while modular technologies ensure that the AI system will continue to evolve to grow and change as needed in a city (Kondor et al., 2024).

But in AI adoption, another challenge faced by inter- and intra-agency coordination includes many city departments such as transportation, health, and energy. Most of the time, each one of them works like an island, resulting in some inefficiency. Setting up a central AI governance structure and creating interagency task forces are various ways that cities come up with ways to handle their projects in a holistic approach. "A monthly coordination meeting helps different departments to align and work collaboratively with each other for proper implementation" (Froehlich et al., 2024).

One of the challenges to this equitable deployment of AI technologies is social and economic inequalities. AI systems run the risk of benefiting the wealthier districts first, leaving behind underprivileged areas. In response, Warsaw, Prague, and Budapest are focused on ensuring that AI benefits are equitably distributed across all city districts. This also involves dedicated AI initiatives aimed at reducing inequity in disadvantaged areas and constant monitoring to identify and resolve any possible social or economic inequalities emerging from the adoption of AI (Pradeep et al., 2025).

Table 9.3 summarizes the principal challenges the Central European Smart Cities of Warsaw, Prague, and Budapest face with regard to the adoption of AI technologies, along with the methods each city deploys in order to try to overcome these challenges.

9.3 COMPARATIVE ANALYSIS WITH OTHER REGIONS

AI has now begun to play a decisive role in bringing a sea of change in the face of urban centers in all parts of the world, mainly through their applications related to making cities smarter. The rationale behind such activity is that it enhances resource management and optimizes the living standards of people for further sustainable development of the community. It might be instructive to compare and contrast the practices of integrating AI in Smart Cities in Central Europe with those of North America, Asia, and Latin America. The approaches taken by each are unique, shaped by their economic, technological, and social contexts, as well as their varied stages of urban development and policy priorities.

Characteristically, AI adoption in Central European intelligent cities is marked by a strong focus on sustainability, which sets a wider agenda for the region, while at the same time, there are very strict attitudes toward data privacy and a rather balanced view on technological progress. The use of AI by Prague and Warsaw has become increasingly wide for improving urban mobility and attaining higher energy efficiency, and also for environmental monitoring. Central European cities can boast of robust data protection legislation that shapes the way AI solutions are brought into service—very often under considerable degrees of regulatory oversight. In Central Europe, there is an emphasis

TABLE 9.3 Key Challenges in AI Implementation in Smart Cities

Challenge	Description	Methods of Overcoming
Data privacy and security	Most of the AI applications for Smart Cities are pretty data-intensive; personal data is a huge constituent of this. Ensuring this is where GDPR compliance assures the privacy of the city's denizens as the mainstay.	- Implementation of data protection policies - Investments in cybersecurity infrastructure - Awareness campaigns about data privacy - Collaboration with EU bodies for guidance
Technological infrastructure	The digital and technological infrastructure is, to say the least, mostly underdeveloped in requirement for higher-order AI systems and platforms in vogue, such as 5G networks and high-broadband networks.	- Upgrading communication networks, including fiber optics and 5G networks - Public-private partnerships to invest in smart infrastructure - Gradual deployment of AI systems in phases to monitor and improve technical readiness
Lack of skilled workforce	AI is a field of high professionalism. There are not enough experts in the fields of data science and machine learning for AI development either in Central European cities.	- Collaboration with universities and research institutions to promote AI education - Creation of specialized training programs - Incentives for international talent and AI startups to establish in these cities
Financial constraints	Significant financial investments in Smart Cities are necessary to implement AI in the cities, which may stand in the way of far-reaching projects due to their limited public budgets	- Securing EU funding for innovation projects - Public-private partnerships to co-fund AI initiatives - Leveraging investment in other sectors like transportation and urban planning to indirectly support AI integration
Integration with existing infrastructure	Lots of Smart City AI solutions have to be integrated into the already existing, very often old infrastructure, and this might be technologically challenging and costly.	- Gradual integration of AI systems with legacy infrastructure - Pilot projects to test AI solutions on a small scale - Use of flexible and modular AI systems that can adapt to existing infrastructure
Public resistance to AI	Skepticism in the acceptance of AI technologies into the day-to-day running of city life is also brought about by the perceived threat to privacy, jobs, and societal norms.	- The implementation of public consultations and awareness campaigns - Showing to public successful AI implementations in other cities - Involving citizens in the development process to increase the people trust in AI solutions
Regulatory and legal challenges	Generally, vagueness or inconsistency in the regulatory framework slows down the processes relating to AI technologies. Most of the AI systems face legal uncertainties with regard to liability, ethics, and decision-making.	- Collaboration with EU institutions to align local regulations with EU guidelines - Establishment of city-specific AI task forces to create clear regulations - Development of ethical guidelines for AI use in public services
Scalability and adaptability of solutions	Nowadays, the biggest challenge in Smart Cities has been the deployment of AI systems that can scale from the entire city and still be adaptable under variable conditions.	- Pilot programs to test scalability of the AI solution before full implementation - Development of AI systems that can grow accordingly with the city's needs - Investment in modular, scalable AI technologies

(*Continued*)

TABLE 9.3 (Continued)

Challenge	Description	Methods of Overcoming
Inter-agency collaboration	AI projects require coordination from a city's departments, which is essentially fragmented, hence making the processes inefficient.	- Establishment of centralized AI governance structures - Regular coordination of meetings between particular city departments - Creation of inter-agency task forces to manage AI projects in the city
Social and economic inequality	Implementation may become unwittingly unfair, the wealthy districts would therefore be the first to move ahead and leave others lagging behind.	- Focus on equitable distribution of AI benefits across all city districts - Tailored AI projects aimed at reducing inequalities in underserved areas - Monitoring and addressing potential social and economic disparities in AI implementations

Source: Author's own work on basis: (Standards for Responsible Use, 2023; Method, 2023; Construction, 2023; National Debut, 2024; State of Warsaw, 2024).

on the collaboration of the public sector, research institutions, and private enterprise in driving innovation. There does exist, however, a very unique context that has been created by means of dependency on EU funding and policies that often turn initiatives on Smart Cities in line with Europe-wide objectives, such as carbon neutrality and energy transition.

North America took another route in Smart City projects with a central focus on innovation and scalability. Cities like Toronto, New York, and San Francisco were fast enough to integrate AI-driven applications into different thematic areas, such as road congestion, public safety, and health management. The high visibility of the tech industry in the regional economy also implies that large-scale private sector investment has tended to assume a shaping role in the development of the sector. While North America is one of those heavily investing in Smart City projects, it sometimes characterizes this approach as allowing the rapid, if not often unconditional, advancement of economic growth and efficiency over concerns related to the protection of data privacy and control. This often involves the deployment of AI in such cities to make urban living more livable, connected, and efficient; it is also equally linked with important ethical issues of AI, including increased surveillance technologies and a likely rise in social inequality (Verma, 2024; Mcneil, 2024; The New York City, 2023; Gourani, 2024).

In Asia, AI for Smart Cities is mostly driven by fast urbanization and a need felt by many to come up with large-scale solutions for managing the effects of densely populated urban areas. Cities like Singapore, Seoul, and Beijing are already pursuing very ambitious AI projects aimed at integrating smart technology in all aspects of urban life, from transportation and energy consumption to waste management and public administration. Most Asian cities underline this aspect by underlining how AI fits into the infrastructural ecosystem of urban living: streamlining road traffic, enhancing public transport, and monitoring air quality. In that respect, AI is often part of a much larger initiative led by the government, in which state involvement and top-down planning are key.

Developments of this sort are also clouded by significant issues from the application of AI as well as questions about the role of privacy and control by the state, the latter very much within an explicit Chinese context. When it comes to efficiency in innovation, coupled with pressures that stem from urban growth, many experts judge Asia as probably the most advanced region currently developing Smart Cities (Nayak, 2024; Trueman and Lago, 2019; Seoul, 2024; The Rise, 2024; Wilan, 2024).

The barriers also come from the fact that the rate of urbanization is extremely high, along with its associated companions like poverty, lack of infrastructure, and other similar elements. AI initiatives in cities like São Paulo, Mexico City, and Buenos Aires are on the rise, but they seem more practical in attempting to solve very practical issues, such as developing safe spaces for citizens, management of waste, or better ways to improve traffic congestion. Compared to North America and Asia, Latin America may experience a relatively slower rate of adoption due to the existence of financial barriers, limited access to state-of-the-art technologies, and less developed regulatory regimes. Efforts should be mentioned in the region to integrate in a more inclusive, social, and people-oriented way. Most of the Smart City projects are in a position to express international cooperation with technology providers for the implementation of AI toward the increased quality of the living level of inhabitants (Brandao, 2024; AI Procurement, 2022; Gomez, 2024; Pozo and Arrojo, 2024; Norak, 2024; Guilera, 2023).

BIBLIOGRAPHY

AI Hungary, (2020). https://ai-hungary.com/en (access data 18.12.2024).

AI Procurement in a Box: Lessons in Best Practice From Brazil, (2022). www.weforum.org/stories/2022/05/the-brazilian-case-for-ai-procurement-in-a-box/ (access data 18.12.2024).

Ali, J., Kumar Singh, S., Jiang, W., Ibrahim Daradkeh, Y., and Mehmood, A. (2025). A deep dive into cybersecurity solutions for AI-driven IoT-enabled smart cities in advanced communication networks. *Computer Communications*, 229, 108000.

Balashova, S.A., and Reshetnikova, M.S. (2024). Business sector and the smart city in the artificial intelligence (AI) era: Ethics, privacy, and cybersecurity. In: A. Visvizi, O. Troisi, & M. Grimaldi (Eds.), *ICT, the Business Sector and Smart Cities*. Abingdon, Oxfordshire, United Kingdom: Routledge, pp. 210–226.

Beijing Municipal Action Plan to Promote "AI+" (2024–2025), (2024). https://cset.georgetown.edu/publication/beijing-ai-plan-2024-2025/ (access data 18.12.2024).

Brader A. (2024). Budapest University Launches AI Training Programme for Business and Finance, www.hungarianconservative.com/articles/current/budapest_metropolitan_university_course_artificial_intelligence/?utm_source=chatgpt.com (access data 18.12.2024).

Brandao, R. (2024). The current scenario of Artificial Intelligence development in Brazil. *Internet Sectoral Overview*, 16(1), 1–52, https://cetic.br/media/docs/publicacoes/6/20240514085413/iso-year-xvi-n-1-ia-development-in-brazil.pdf (access data 18.12.2024).

Budapest to Host New Research Hub for AI-driven Cancer Diagnostics, (2023). https://hipa.hu/news/budapest-to-host-new-research-hub-for-ai-driven-cancer-diagnostics/?utm_source=chatgpt.com (access data 18.12.2024).

Budapest's AI-Powered Personalized Travel Planning Experience, (2023). www.thinkdigital.travel/x-awards-entries/budapests-ai-powered-personalized-travel-planning-experience (access data 18.12.2024).

Construction of the Artificial Intelligence Research Centre in Warsaw, (2023). https://opi.org.pl/en/construction-of-the-artificial-intelligence-research-centre-in-warsaw/?utm_source=chatgpt.com (access data 18.12.2024).

Continental Will Develop Its AI-Based Autonomous Driving Systems in Budapest, (2018). https://hipa.hu/news/continental-will-develop-its-ai-based-autonomous-driving-systems-in-budapest/?utm_source=chatgpt.com (access data 18.12.2024).

Empowering Humanity to Shape Its Future with AI, (2024). www.aic.fel.cvut.cz/?utm_source=chatgpt.com (access data 18.12.2024).

Estebsari, A., and Werna, E. (2022). *Smart Cities, Grids, Homes and the Workforce: Challenges and Prospects, Current State of Art in Artificial Intelligence and Ubiquitous Cities.* Singapore: Springer, pp. 17–39.

Experience the Future of AI in Budapest, (2024). https://budappest.com/budapest-events/experience-the-future-of-ai-in-budapest/ (access data 18.12.2024).

Froehlich, J.E., Li, C., Hosseini, M., Sevtsuk, A., and Eisenberg, Y. (2024). The future of urban accessibility: The role of AI, ASSETS 2024. *Proceedings of the 26th International ACM SIGACCESS Conference on Computers and Accessibility*, 135.

Gomez, D. (2024). AI Industry in Mexico 2024, https://alcor-bpo.com/ai-mexico/ (access data 18.12.2024).

Gourani, S. (2024). San Francisco's Edge in the AI Race, www.forbes.com/sites/soulaimagourani/2024/02/25/san-franciscos-edge-in-the-ai-race-the-role-of-education-and-inclusivity/ (access data 18.12.2024).

Guilera, S. (2023). Localizing Urban AI: Developments in Latin America, https://medium.com/urban-ai/localizing-urban-ai-developments-in-latin-america-6b402eef4ab2 (access data 18.12.2024).

Hungary Adopts New AI Strategy, (2020). https://cms-lawnow.com/en/ealerts/2020/09/hungary-adopts-new-ai-strategy?utm_source=chatgpt.com (access data 18.12.2024).

Hungary's Artificial Intelligence Strategy 2020-2030, (2020). Ministry for innovation and Technology, https://ai-hungary.com/api/v1/companies/15/files/146074/view (access data 18.12.2024).

Jonek-Kowalska, I., and Wolniak, R. (2023). *Towards Sustainability and a Better Quality of Life?* Routledge: London, UK.

Jonek-Kowalska, I., and Wolniak, R. (2025). *Exclusion in Smart Cities Assessment and Strategy for Strengthening Resilience.* Routledge: London, UK.

Kolisnychenko, A. (2023). New Center of Excellence in Prague Helps Companies with Generative AI, https://praguemorning.cz/new-center-of-excellence-in-prague-helps-companies-with-generative-ai/?utm_source=chatgpt.com (access data 18.12.2024).

Kondor, D., Hafez, V., Shankar, S., Wazir, R., and Karimi, F. (2024). Complex systems perspective in assessing risks in artificial intelligence, *Philosophical Transactions of the Royal Society A: Mathematical, Physical and Engineering Sciences*, 382(2285), 20240109.

Krasavina, A. (2023). Czech Republic – National Artificial Intelligence Strategy of the Czech Republic, https://digital-skills-jobs.europa.eu/en/actions/national-initiatives/national-strategies/czech-republic-national-artificial-intelligence?utm_source=chatgpt.com (access data 18.12.2024).

Mcneil, J. (2024). Smart Cities: Where Human and Artificial Intelligence Meet, https://torontoglobal.ca/technology/smart-cities-where-human-and-artifical-intelligence-meet-2024/ (access data 18.12.2024).

Method for Designing AI Standards for Warsaw, (2023). https://um.warszawa.pl/waw/ai-guidelines/-/method-for-designing-ai-standards-for-warsaw-1 (access data 18.12.2024).

National Artificial Intelligence Strategy of the Czech Republic, Ministry of Industry and Trade of the Czech Republic, (2024). www.mpo.gov.cz/assets/en/guidepost/for-the-media/press-releases/2019/5/NAIS_eng_web.pdf (access data 18.12.2024).

National Debut of the State of Warsaw AI Report at Meetup65 in Warsaw, (2024). https://digitalpoland.org/en/blog/2024/11/national-debut-of-the-state-of-warsaw-ai-report-at-meetup65-in-warsaw?utm_source=chatgpt.com (access data 18.12.2024).

Nayak, R. (2024). Singapore's forward-thinking approach to AI regulation, www.diligent.com/resources/blog/Singapore-AI-regulation (access data 18.12.2024).

Norak, M. (2024). Argentina's plan to fight crime with AI draws concerns from rights groups, www.biometricupdate.com/202408/argentinas-plan-to-fight-crime-with-ai-draws-concerns-from-rights-groups (access data 18.12.2024).

Obaideen, K., Albasha, L., Iqbal, U., and Mir, H. (2024). Wireless power transfer: Applications, challenges, barriers, and the role of AI in achieving sustainable development goals – A bibliometric analysis. *Energy Strategy Reviews*, 53, 101376.

Ojha, N., and Nikhil, V.P. (2024). Challenges of AI implementation for boosting circular economy in smart cities. In: V. Kandpal, E.D.R. Santibañez-Gonzalez, & P. Chatterjee (Eds.), *Smart Cities and Circular Economy: The Future of Sustainable Urban Development.* Bingley, United Kingdom: Emerald Publishing, pp. 215–233.

Pesti, D. (2024). Budapest Implements AI to Determine New Speed Limits to Improve Road Safety. https://budappest.com/budapest-implements-ai-to-determine-new-speed-limits-to-improve-road-safety/?utm_source=chatgpt.com (access data 18.12.2024).

Pozo, C., and Arrojo, D.R. (2024). Mexico's Bet on Artificial Intelligence, www.wilsoncenter.org/blog-post/mexicos-bet-artificial-intelligence (access data 18.12.2024).

Pradeep, M., Kumari, S., Tyagi, A.K., and Tiwari, S. (2025). Smart hospital in smart cities. In: A.K. Tyagi (Ed.), *Digital Twin and Blockchain for Smart Cities*. Hoboken, New Jersey, USA: Wiley, pp. 579–603.

Revolutionizing Urban Mobility: BKV Zrt.'s AI-Driven Solutions for Budapest's Transport, (2024). www.cash-platform.com/revolutionizing-urban-mobility-bkv-zrt-s-ai-driven-solutions-for-budapests-transport/?utm_source=chatgpt.com (access data 18.12.2024).

Seoul Aims to Become the Best City in AI Utilization; AI-Based Administration to Enhance Well-being of Citizens, (2024). https://english.seoul.go.kr/seoul-aims-to-become-the-best-city-in-ai-utilization-ai-based-administration-to-enhance-well-being-of-citizens/ (access data 18.12.2024).

Singh, B., and Nayyar, A. (2024). Exploring diverse use cases of digital twins projecting digital transformation: Unlocking potential, addressing challenges and viable solutions. In: *Digital Twins for Smart Cities and Villages*. Amsterdam, Netherlands: Elsevier, pp. 631–655.

Smart Budapest, (2024). www.thesmartcityjournal.com/en/cities/smart-budapest (access data 18.12.2024).

Standards for Responsible Use of Generative Artificial Intelligence in the City of Warsaw, (2023). https://um.warszawa.pl/documents/129814944/0/ENG_Guidelines+for+Responsible+Use+of+GenAI_250303.pdf/3a194d7e-cb69-1312-1879-b68b2bc22d86?t=1742812197368 (access data 18.12.2024).

State of Warsaw AI, Report on the Artificial Intelligence Sector in Warsaw, (2024). City Hall of Warsaw.

Stofan, D. (2024). Prague's AI Traffic Fix: 5 Key Takeaways From Smart Mobility Day, https://blog.goodvisionlive.com/smart-mobility-prague (access data 18.12.2024).

Sztuczna inteligencja na rzecz miasta, (2023). https://um.warszawa.pl/-/sztuczna-inteligencja-na-rzecz-miasta?utm_source=chatgpt.com (access data 18.12.2024).

Sztuczna inteligencja w Warszawie, (2024). https://um.warszawa.pl/-/sztuczna-inteligencja-w-warszawie?utm_source=chatgpt.com (access data 18.12.2024).

The New York City Artificial Intelligence Action Plan, (2023). www.nyc.gov/assets/oti/downloads/pdf/reports/artificial-intelligence-action-plan.pdf (access data 18.12.2024).

The Rise of AI in South Korea: Transforming Industries and Shaping the Future, (2024). www.newnop.com/blog/the-rise-of-ai-in-south-korea (access data 18.12.2024).

Transforming Prague into a Significant European AI Hub, (2023). https://prg.ai/en/?utm_source=chatgpt.com (access data 18.12.2024).

Trueman, Ch., and Lago, Ch. (2019). How Singapore is using artificial intelligence, www.cio.com/article/221994/how-singapore-is-using-artificial-intelligence.html (access data 18.12.2024).

Verma, P. (2024). Shaping the Future in Toronto: AI and Tech Innovation in Canada's Silicon Valley, https://sector-pulse.com/sectorinsights/shaping-the-future-in-toronto-ai-and-tech-innovation-in-canadas-silicon-valley, (access data 18.12.2024).

Warsaw AI, (2024). https://warsaw.ai/ (access data 18.12.2024).

Warszawa – Smart City, (2019). https://um.warszawa.pl/-/warszawa-smart-city?utm_source=chatgpt.com (access data 18.12.2024).

Weber, H. (2024). Czech Republic Unveils Ambitious AI Strategy for 2030, www.pragueforum.cz/czech-republic-unveils-ambitious-ai-strategy-for-2030/?utm_source=chatgpt.com (access data 18.12.2024).

Wilan, Z. (2024). Beijing launches AI public platform as demand mounts for computing power, www.globaltimes.cn/page/202404/1310215.shtml (access data 18.12.2024).

Wolniak, R., and Stecuła, K. (2024). Artificial intelligence in smart cities—applications, barriers and future directions: A review. *Smart Cities*, 7, 1346–1389.

CHAPTER 10

AI as a Catalyst for Innovation

10.1. AI-DRIVEN AUTOMATION AND OPTIMIZATION IN URBAN MANAGEMENT

In an era of dynamic social, environmental, and technological changes, cities need innovative solutions that will enable them to function and develop effectively. Artificial Intelligence is becoming not only a tool for optimizing and automating urban processes, but above all, a catalyst for innovation. AI can help in the creation of new models of management, services, interactions, and decision-making. Traditional methods of urban space management often are insufficient in the context of growing populations and the complexity of infrastructure systems. What is more, the meaning of sustainable development cannot be forgotten. In response to these challenges, AI is introducing innovative approaches based on data, prediction, and adaptation. It can transform the way cities are managed—from daily operations to strategic planning in cities.

In the past, maintaining urban infrastructure was done through routine inspections or responding to emergencies. Thanks to Artificial Intelligence, cities can now create intelligent monitoring systems that analyze current data from sensors, cameras, and IoT devices (Unciano and Khan, 2025). These systems detect patterns that indicate faults that may occur and allow for predictive infrastructure maintenance—before a failure occurs. This innovative approach can help, for example, minimize power outages, water main breaks, and road surface damage.

Energy management in cities is also going through transformation thanks to AI. Innovative systems called smart grids are becoming more and more popular and common in cities (Omitaomu and Niu, 2021; Onen, 2022). They allow not only for monitoring energy consumption but also for dynamically adjusting supply to demand. Learning

DOI: 10.1201/9781003587613-11

algorithms analyze data on weather conditions, residents' lifestyles, and production from renewable sources in real time. Based on this vast amount of data, AI can intelligently control the supply of electricity, heating, and cooling (Qi et al., 2024). Furthermore, AI also enables the implementation of innovative prosumer models (Hua et al., 2022). In such models, residents not only consume but also produce energy. Automatic management of excess energy and balancing local micro-grids are some of the characteristic features of smart buildings/smart homes in Smart Cities (Jiang and Fei, 2014). In the future, this could lead to the development of self-regulating energy systems that can learn and adapt to changes in real time. This can help with energy and cost savings.

Another AI-powered innovation is the concept of Mobility-as-a-Service (MaaS), which is also popular in Smart Cities (Maas, 2022). It is an integrated transportation service that combines various modes of transportation (buses, city bikes, scooters, taxis, and metro) into a single platform. AI can analyze various data, including location data, weather conditions, user preferences, and the availability of specific modes of transportation. Based on this data, AI can suggest the most efficient door-to-door route. In larger cities, residents can use AI-powered applications that not only plan trips but also dynamically modify them in real time, taking into account traffic jams, delays, and breakdowns. Thanks to AI, cities can predict congestion, optimize schedules, and reduce emissions. In the literature, there is, for example, a MaaS platform in rural areas of Southern Minnesota (Khani, 2024), which integrates interactive trip planning, mobile payment, and demand response transit (DRT) booking features. The aim of this platform is to develop the rural transit user experience and encourage more use of transit. There are also examples of Maas in Helsinki (Audouin and Finger, 2018), Paris (Reyes Madrigal et al., 2023), Lisbon (Florêncio, 2021), and more cities.

AI not only automates operations but also helps in the implementation of new models for sustainable resource management. One such model is the circular economy (Geisendorf and Pietrulla, 2018). The traditional linear approach (based on production, consumption, and disposal) nowadays can be perceived as inefficient and unsustainable, especially when it comes to urbanization and limited resources. Thanks to AI, intelligent management of the lifecycle of products and raw materials is possible. This helps to reduce waste and increase resource efficiency. For example, AI systems analyze data from sensors in waste containers (so-called smart bins), which monitor their fill levels in real time (Haque et al., 2020). Algorithms use this information to optimize waste collection vehicle routes. This not only reduces CO_2 emissions and fuel consumption but also logistics costs. Then, garbage trucks do not have to run on a set schedule. They can operate in a dynamic mode, for example, only where containers need (indeed) to be emptied. AI can also categorize waste based on data from cameras and vision systems in sorting plants (Bhattacharjee and Bhave, 2025). Due to machine learning, it is possible to automatically recognize material type—for example this can be PET plastic, glass, metal, etc. In practice, then less waste goes to landfills and recovery of valuable raw materials is greater (they can be reused in production). Additionally, AI can be used to model waste flows in a given city or district, allowing officials to predict when and where the highest demand will be

TABLE 10.1 Examples of AI Usage in Urban Management—Innovations

Area	Innovative Usage of AI	Benefits and Effects
Infrastructure management	Predictive maintenance based on data from sensors, cameras, and IoT devices	Early fault detection, fewer breakdowns, reduced repair and maintenance costs
Energy and utilities management	Smart grids, analysis of energy consumption and production, automated management of prosumer microgrids	Energy optimization, reduced losses, integration of renewables, support for sustainable urban development
Transport and mobility	Mobility-as-a-Service platforms, dynamic route planning, integration of multiple transport modes	Reduced congestion, more convenient commuting, better infrastructure planning
Waste management	Analysis of smart bin data, route optimization, automated waste sorting using AI vision systems	Lower CO_2 emissions, reduced logistics costs, higher recycling efficiency, increased recovery of raw materials
Crisis and climate management	Modeling disaster scenarios, predicting threats (e.g., floods), AI-assisted evacuation and response planning	Faster and more precise crisis response, improved safety and resilience of the urban population
Urban and spatial planning	AI analysis of demographic, environmental, and transportation data to optimize land use and urban layout	Improved planning decisions, higher quality of life, promotion of balanced and sustainable city development

Source: Authors' own work.

on the system. Based on historical data and seasonal factors (e.g., holidays, festivals, weather conditions), AI can support decisions about the deployment of additional containers in specific locations. For example, in some countries, holidays when people visit cemeteries in large numbers (to light candles for the deceased) would require such measures.

In the context of climate change and the risk of natural disasters, AI is becoming a catalyst for innovative crisis management methods. AI enables not only rapid response but also modeling threat scenarios (e.g., floods, earthquakes, pandemics). Predictive systems can suggest preventative measures such as evacuations, route changes, and real-time resource distribution (Rehan, 2022). Table 10.1 summarizes the above presented examples.

10.2 ENHANCING PUBLIC SERVICES THROUGH AI INNOVATIONS

Modern cities face growing expectations from their residents regarding the quality, speed, and availability of public services. On the one hand, we are dealing with an increasingly diverse population with diverse needs. This is due, in part, to technological progress and digitization, which allows some social groups (usually younger, but not exclusively) to make excellent use of new technologies, media, and IT tools. Other groups, on the other hand, face challenges and barriers to using technological innovations. Unfortunately, there is even talk of a digital divide. On the other hand, these differences stem from limited human and financial resources within the administration. In this context, Artificial Intelligence not only automates existing processes but also introduces innovations that have the potential to significantly change the way public services are delivered. AI becomes a tool for transformation—both in terms of efficiency, personalization, accessibility, and proactive management of residents' needs.

One of the areas of AI use in public services is administrative services. Traditional bureaucratic processes can be time-consuming and complex. This often discourages citizens from using them. Dealing with a matter, for example, in an office often requires significant time and complex procedures. In this area, AI has significant potential to improve current situations. One example is virtual assistants. They can take the form of chatbots based on natural language processing. Such a solution can positively impact the automation of many interactions. AI can schedule office appointments, check application status, or explain regulations. Moreover, such systems are available 24/7 and can support multiple languages, increasing accessibility for migrants and those with less digital literacy. AI can also translate difficult legal language into terms understandable for the average user. In the literature, you can already find articles that discuss chat bots in official environments, for example "*Super AI, generative AI, narrow AI and chatbots: An assessment of artificial intelligence technologies for the public sector and public administration*" (Damar et al., 2024), "*How AI-Chatbots can make Dubai smarter*?" (Mohasses, 2019), "*Public value positions and design preferences toward AI-based chatbots in e-government. Evidence from a conjoint experiment with citizens and municipal front desk officers*" (Hemesath and Tepe, 2024).

Healthcare systems are another area where AI is bringing significant innovation. AI-based platforms enable the analysis of health data, enabling them to provide personalized recommendations and predict disease risk. They can also manage access to specialists. Patients can receive reminders for preventive screenings, monitor their parameters using wearable devices, and receive automatic analysis of results (Mukhopadhyay et al., 2022). AI can also support diagnosis—for example, systems that analyze medical images (e.g., X-rays, MRIs), identifying pathological changes faster than a doctor. Such solutions contribute to reducing the burden on healthcare services and increasing their efficiency.

In the context of public education, AI creates opportunities to create new study programs tailored for given students and introduce new ways of learning. Machine learning can analyze students' progress and recommend educational materials tailored to their level and learning style. Students with learning difficulties can receive additional support, while gifted ones can receive more advanced content. AI also enables automated test grading, quiz generation, and teacher support. Examples include digital learning platforms with personalized learning paths and interactive materials (Bibri and Jagatheesaperumal, 2023).

City management using AI can also include optimizing communication between city hall and residents. By analyzing data from various sources (such as social media, online forms, and other resident reports), AI-based systems can quickly identify problems. For example, if numerous residents report a lighting problem in a given district, the system can automatically generate a request for technical assistance.

Another potential area of innovative AI applications is road traffic management and public transportation. The first example can be intelligent traffic monitoring systems. They can have different functions, for instance predicting traffic jams, automatically managing traffic lights, and adjusting public transportation schedules. AI can integrate data on pedestrians, bicycles, and vehicle traffic to recommend optimal routes and travel times. Furthermore, it can also forecast transportation needs (for example, higher traffic during mass events or holidays) and prepare infrastructure for them.

It is also worth mentioning social services, which, with the help of AI, can better adjust assistance to the needs of a given resident. Due to the analysis of demographic, economic, and health data, AI systems can automatically qualify people for assistance programs. They can also propose forms of support and monitor the assistance. This allows people at risk of social exclusion to receive assistance more quickly and allows offices to manage public funds more effectively.

Finally, AI can play a significant role in increasing the accessibility of services. This refers to different groups, for example seniors, people with disabilities, and residents of digitally excluded areas. Thanks to features such as speech synthesizers, senior-friendly interfaces, and automatic translations, public services have become more inclusive. Table 10.2 summarizes the above presented examples.

The use of Artificial Intelligence in the public services sector offers enormous innovative potential. AI not only enables the automation of certain processes but also makes them faster, personalized, and more efficient. AI-powered innovations have significant potential to improve the quality of life for city residents, especially those previously at risk of digital, social, or health exclusion. AI can provide better data management, more accurate medical diagnoses, adjusted education, and easier communication with government agencies. This allows cities to operate more efficiently. Smart Cities can also more consciously respond to the diverse needs of the citizens. Innovations can support sustainable development which nowadays is so important. They should also contribute to greater social justice, the efficient use of public resources, and the activation of local communities. In this role, AI becomes not only a technological tool but also a social one.

However, despite these benefits, there are many challenges and threats. First, there are ethical issues—concerning data privacy, algorithm transparency, and accountability for AI-made decisions. Automatic eligibility for social assistance and health status analysis are very sensitive areas that must be protected. Second, the development of AI in public

TABLE 10.2 Examples of Innovative AI Usage in Public Services

Sector	AI Application	Benefits
Administration	Chatbots, automated forms, natural language processing	24/7 availability, reduced queues, simplified administrative language
Healthcare	Early diagnosis, health recommendations, wearable data analysis	Prevention, reduced treatment costs, faster assistance
Education	Personalized materials, automated testing, AI-powered learning assistants	Equal opportunities, better learning customization
Transport	Smart timetables, traffic management, demand forecasting	Shorter travel time, less pollution, improved mobility
Social Services	Needs analysis, rapid qualification for support, monitoring aid effectiveness	Better benefit matching, faster intervention
Inclusivity	Content translation, voice assistants, simplified user interfaces	Greater accessibility of services, independence for people with disabilities

Source: Authors' own work.

services can deepen citizens' dependence on technology and contribute to the weakening of traditional social relationships, limiting direct contact with other people. Cooperation with another human requires effort and often compromise. In conversations, AI adapts to the given person and meets their expectations. This can be addictive and may lead to difficulties in relationships with other people in the future. People accustomed to giving orders, for example, to a chatbot, will be reluctant to work with another human who has their own opinions, ideas, and views. In the long run, this could lead to a change in human behavior within a group, leading to a greater willingness to interact with AI than with humans. Lifestyles are also changing—we are increasingly shifting decisions and actions to digital systems, which carries the risk of losing independence, patience, and sometimes the ability to think critically.

For this reason, it is crucial to implement AI responsibly, respecting social values, taking into account the diversity of residents, and maintaining human control over the system. Technologies can support the common good, but only if they are designed in a transparent, equitable, and human-centered manner. In summary, AI has the potential to significantly improve public services, making them more accessible and efficient. This represents a great opportunity for cities of the future—but also a responsibility to develop these technologies ethically, wisely, and sustainably.

10.3 AI IN URBAN PLANNING AND DEVELOPMENT

Modern cities are facing the increased complexity of spatial, social, and economic processes. At a time of fast-growing urbanization, growing populations, and problems associated with climate, mobility, access to services, and governance of public space, traditional urban planning methods typically fail. Artificial Intelligence in this context can serve as a vehicle to support planning processes, make data analysis easier, and help evolve more intelligent, responsive, and sustainable cities.

Urban AI is no longer something that will happen in the future but is already a solution to the need for faster and more holistic decision-making for the city space. AI provides the capability for instant evaluation of massive amounts of spatial, demographic, environmental, and infrastructure data (Soomro et al., 2019). This, in turn, when executed properly, can significantly accelerate and improve planning processes.

One of the most important applications of AI in urban planning is the analysis and prediction of spatial development. Traditional planning has a tendency to rely on the historical data, maps, and planning documents. Meanwhile, AI has the potential to process satellite data (Ghiglione and Serra, 2022), IoT sensor data (Mukhopadhyay et al., 2021), drone data (Kim et al., 2021), and land survey data. AI can disclose patterns and trends from this data. For example, these can be suburban expansion, shifts in building density, the risk of illegal development, or decline of open areas. In this manner, spatial planning is made more responsive, proactive, and attuned to the actual needs of citizens.

AI also enables modeling city situations, simulating various space development scenarios. For example, while planning a new residential area, the AI system is able to simulate its impact on traffic, public services demand (schools, clinics), environmental impact, and air quality. This enables forecasting the impact of urban planning before implementation. It minimizes the chances of planning errors and makes city investments more effective.

AI can also support the development of sustainable cities by identifying areas requiring revitalization (Cao et al., 2025). Based on data on energy consumption, the technical condition of buildings, noise levels, air pollution, and population density, AI can pinpoint neighborhoods with a reduced quality of life. City authorities can then focus investments in social, transportation, and energy infrastructure more precisely and equitably.

Another important area is green infrastructure planning. AI can analyze spaces for the presence of green areas, their accessibility to residents, the quality of vegetation, and their impact on the urban microclimate (Patil et al., 2024). This allows maps of so-called green deserts to be created—areas lacking sufficient greenery and therefore susceptible to urban heat island phenomena. As a result, AI supports sustainable and resilient cities that better cope with climate change and provide better living conditions for their residents.

Public participation is also becoming increasingly important in urban development planning, and AI can play a very positive role here. AI-based tools can analyze residents' opinions expressed on social media, online surveys, city forums, or during public consultations. Based on this, maps of "public sentiment" and lists of the most frequently expressed needs and concerns are created. This enables more democratic and transparent spatial planning that takes into account the opinions of citizens, not just the assumptions of officials or developers.

Another application of AI is the analysis of data related to mobility and accessibility. Machine learning can analyze data from GPS, ticketing systems, transportation apps, and traffic lights. Based on this, AI can identify so-called transportation white spots—places with poor connectivity and requiring intervention in the form of a new bus line, bikeshare station, or improved pedestrian infrastructure. It can also support the planning of public transportation networks to make them more accessible, punctual, and environmentally friendly.

It is also worth emphasizing that AI can assist city planners in managing land and spatial resources (Patil et al., 2024). By analyzing data on land ownership, land use, development plans, as well as geotechnical and environmental data, AI systems can identify the best locations for new investments—both public and private. This approach increases the transparency and efficiency of investment processes and reduces the risk of spatial conflicts.

Last but not least, it must be highlighted that the development of AI in urban planning also presents challenges. One of these is access to data—its quality, timeliness, and openness. Another is the issue of accountability. The question is: *who is responsible for decisions made based on AI recommendations*? This issue is one of the most important and difficult to resolve. Finally, ethics matter is essential—algorithms can reinforce inequalities if they learn from biased data. Therefore, Artificial Intelligence can reduce inequalities between people, but it can also increase them. On the one hand, Smart Cities are associated with inclusiveness. But on the other hand, the problem of exclusion may also be intensified as a side effect. As a conclusion, Artificial Intelligence is a useful tool in urban development and planning. With its application, it is possible to analyze data more effectively and predict the outcome of decisions. AI can help improve social engagement and support sustainable urban development. It is, nevertheless, necessary to create such tools responsibly. AI-based solutions need to be transparent, ethical, and oriented toward the common good. Then, Smart Cities will be not only more efficient but, above all, more human.

BIBLIOGRAPHY

Audouin, M., and Finger, M. (2018). The development of Mobility-as-a-Service in the Helsinki metropolitan area: A multi-level governance analysis. *Research in Transportation Business & Management*, 27, 24–35.

Bhattacharjee, T., and Bhave, S. (2025). Application of AI-enabled computer vision technology for segregation of industrial plastic wastes. In *Agroindustrial Waste Management and Natural Resources Conservation* (pp. 233–269). Apple Academic Press.

Bibri, S.E., and Jagatheesaperumal, S.K. (2023). Harnessing the potential of the metaverse and artificial intelligence for the internet of city things: Cost-effective XReality and synergistic AIoT technologies. In *Smart Cities* (Vol. 6, Issue 5, pp. 2397–2429). https://doi.org/10.3390/smartcities6050109

Cao, J., Tang, H., and Zhou, Y. (2025). Pathways of artificial intelligence applications, financial resilience, and rural revitalization. *Finance Research Letters*, 85, 108017.

Damar, M., Özen, A., Çakmak, Ü.E., Özoğuz, E., and Erenay, F.S. (2024). Super AI, generative AI, narrow AI and chatbots: An assessment of artificial intelligence technologies for the public sector and public administration. *Journal of AI*, 8(1), 83–106.

Florêncio, M.G. (2021). *Diagnose of the Smart City Concept Project Scope Definition and Implementation of Mobility as a Service in Lisbon*. Universidade NOVA de Lisboa (Portugal).

Geisendorf, S., and Pietrulla, F. (2018). The circular economy and circular economic concepts—a literature analysis and redefinition. *Thunderbird International Business Review*, 60(5), 771–782.

Ghiglione, M., and Serra, V. (2022). Opportunities and challenges of AI on satellite processing units. *Proceedings of the 19th ACM International Conference on Computing Frontiers*, 221–224.

Haque, K.F., Zabin, R., Yelamarthi, K., Yanambaka, P., and Abdelgawad, A. (2020). An IoT based efficient waste collection system with smart bins. *2020 IEEE 6th World Forum on Internet of Things* (WF-IoT), 1–5.

Hemesath, S., and Tepe, M. (2024). Public value positions and design preferences toward AI-based chatbots in e-government. Evidence from a conjoint experiment with citizens and municipal front desk officers. *Government Information Quarterly*, 41(4), 101985.

Hua, W., Chen, Y., Qadrdan, M., Jiang, J., Sun, H., and Wu, J. (2022). Applications of blockchain and artificial intelligence technologies for enabling prosumers in smart grids: A review. *Renewable and Sustainable Energy Reviews*, 161, 112308.

Jiang, B., and Fei, Y. (2014). Smart home in smart microgrid: A cost-effective energy ecosystem with intelligent hierarchical agents. *IEEE Transactions on Smart Grid*, 6(1), 3–13.

Khani, A. (2024). MnDOT's Mobility-as-a-Service Platform: Assessing User Behavior and Measuring System's Benefits, Center for Transportation Studies, University of Minnesota, Report No. CTS 24-01. https://conservancy.umn.edu/items/763a7c0d-86d4-483e-b8c4-e9199072d204

Kim, B., Jung, J., Min, H., and Heo, J. (2021). Energy efficient and real-time remote sensing in AI-powered drone. *Mobile Information Systems*, 2021(1), 6650053.

Maas, B. (2022). Literature review of mobility as a service. *Sustainability*, 14(14), 8962.

Mohasses, M. (2019). How AI-Chatbots can make Dubai smarter? *2019 Amity International Conference on Artificial Intelligence (AICAI)*, 439–446.

Mukhopadhyay, S.C., Suryadevara, N.K., and Nag, A. (2022). Wearable sensors for healthcare: Fabrication to application. *Sensors*, 22(14), 5137.

Mukhopadhyay, S.C., Tyagi, S.K.S., Suryadevara, N.K., Piuri, V., Scotti, F., and Zeadally, S. (2021). Artificial intelligence-based sensors for next generation IoT applications: A review. *IEEE Sensors Journal*, 21(22), 24920–24932.

Omitaomu, O.A., and Niu, H. (2021). Artificial intelligence techniques in smart grid: A survey. *Smart Cities*, 4(2), 548–568.

Onen, A. (2022). Role of artificial intelligence in smart grids. *Electrical Engineering*, 104(1), 231.

Patil, V., Patil, J., Kadam, A., Patil, A.R., Mokashi, D., and Lonare, G.M. (2024). AI-driven green space optimization for sustainable urban parks: Enhancing biodiversity and resource efficiency. *Library of Progress-Library Science, Information Technology & Computer*, 44(3), 3412–3417.

Qi, H., Ouyang, Q., and Ma, L. (2024). Application of artificial intelligence control in the control system of cooling and heating energy stations. *Thermal Science*, 28(2 Part B), 1321–1328.

Rehan, H. (2022). Enhancing disaster response systems: Predicting and mitigating the impact of natural disasters using AI. *Journal of Artificial Intelligence Research*, 2(1), 501.

Reyes Madrigal, L.M., Nicolaï, I., and Puchinger, J. (2023). Pedestrian mobility in Mobility as a Service (MaaS): Sustainable value potential and policy implications in the Paris region case. *European Transport Research Review*, 15(1), 13.

Soomro, K., Bhutta, M.N.M., Khan, Z., and Tahir, M.A. (2019). Smart city big data analytics: An advanced review. *Wiley Interdisciplinary Reviews: Data Mining and Knowledge Discovery*, 9(5), e1319.

Unciano, N., and Khan, Z. (2025). AI-enabled digital twin framework for predictive maintenance in smart urban infrastructure. *Journal of Smart Infrastructure and Environmental Sustainability*, 2(1), 1–10.

CHAPTER 11

Future Prospects

11.1 THE ETHICAL IMPLICATIONS OF AI IN URBAN ENVIRONMENTS

Artificial Intelligence in city settings is to be bound by a host of ethical considerations highly relevant to be closely looked at. While cities continue ramping up AI in everything from traffic smart systems to predictive policing and urban planning, enormous potential benefits may accrue. The most advanced developments raise important questions concerning fairness, privacy, and accountability, along with its impacts on social equity (Dong and Liu, 2023).

The second major ethical issue is that of privacy. Artificially Intelligent systems need massive amounts of data, which may include sensitive personal information. Data collection and use would likely violate the rights of individuals in an urban environment where surveillance technologies are increasingly applied unless adequate protection mechanisms are established (Phillips and Jiao, 2023). For example, AI applied in the monitoring of public space raises several critical questions, such as to what extent there needs to be tracking of movements and activities of citizens by whom and how the information gained will be used. It risks becoming routine in cities with little transparency or accountability, hence undermining rights to privacy and creating a general atmosphere of distrust among citizens and those in authority (Arora et al., 2023).

The second ethical issue is that these AI algorithms may be biased. Cities are cosmopolitan and hence consist of a large demographic, socio-economic, and cultural groups. Unless highly cautiously designed, the AI system will foster and amplify the inequalities already present in the setting. AI in predictive policing can over-target specific communities, thus reinforcing racial or socioeconomic prejudices (Momen, 2023). More significantly, those are the technologies of Smart Cities that make decisions related to access to

DOI: 10.1201/9781003587613-12

public services or resources where incomplete and biased data relies on such technologies, keeping the marginal groups at a disadvantageous condition always. It is fundamental that the construction of AI systems be by design fair and nondiscriminatory. A prospect might be given to detection, thus correction. Biases can indeed facilitate key leverage for social justice within the context of a city (Schintler and McNeely, 2022; Wolniak and Jonek-Kowalska, 2022).

The second major challenge associated with the use of AI in cities concerns accountability. In a context where these AI systems operate somewhat independently while making decisions without any form of human intervention, question marks are sent out about responsibility in case of some wrongdoing. But in case there is misinterpretation by the AI system and it causes harm—for example, malfunction in the traffic system or unfair distribution of resources—the question of responsibility remains obscure (Helbing et al., 2021). Does it rest with the system developers, the policymakers who implemented it, or is it the AI? Well, clear lines of accountability are required to ensure cities remain accountable to their citizens and that systems can be fixed when they break down (Duan et al., 2022). Also making the AI ethics potentially inextricable at urban centers, therefore, comes through job displacements. Quite understandably, once AI technologies permeate urban transportation, health, and public service, then this would nullify some categories of employments and translate into shrunken economic opportunities in promotion of inequalities. Basic question: "How the support will be extended toward workers who get replaced by automation, or in designing ways whereby benefits accrued through AI should be more equitably distributed." This might widen the existing social rifts and undermine exactly those advantages that AI is supposed to bring (Darmawan et al., 2023; Tao et al., 2022).

Ethical implications of AI in cities cannot be understood outside the broader context of power dynamics (Skubis et al., 2024. One might say, who controls the AI systems? But who makes the decision on their implementation? Very often, development and deployment of AI technologies rest in the hands of one or a few large corporate or government hands. This can raise a number of concerns that such entities may actually employ the use of AI for interests other than the public will. The use of AI does raise another question about democracy and to what extent citizens are allowed to make decisions about their communities. The processes of development and implementation that involve urban AI technologies should be transparent, inclusive, and representative of a variety of stakeholders in such a way that the future of the city will be just and ethical (Sawhney, 2023; Jonek-Kowalska and Wolniak, 2025).

Some aspects related to the ethics of the usage of AI in Smart Cities are reflected in Table 11.1. The above issues underpin the cautious approach to have toward these technologies and for the control of responsible and objective diffusions.

The ethical challenges regarding AI in Smart Cities can be resolved only with a multidimensional approach that foregrounds concerns like fairness, transparency, and accountability prominently, including inclusiveness. Probably, one of the most feasible ways to handle the bias inside an AI system is by using a representative and varied dataset during

TABLE 11.1 Main Ethical Implication of AI Implementation in Smart City

Implication	Description
Bias	AI systems can further exacerbate or even create existing biases—social, racial, or economic—at the expense of already marginalized communities because data is biased or algorithmically designed with bias.
Privacy	Most uses of AI within Smart Cities involve huge volumes of data that breach private information if not well managed or protected. A challenge can also be in terms of the threat of surveillance.
Transparency	AI systems in Smart Cities often act like "black boxes," whose inner mechanisms of decision-making are not that understandable to a layperson, let alone the developers themselves, reducing accountability.
Accountability	Determining who is responsible when AI systems malfunction or cause harm is difficult, especially when decisions are made autonomously by algorithms.
Social equity	This is mainly unequal access to technologies or services in AI, and these further differences are increased, specifically if the marginalized group remains excluded or underserved.
Job displacement	The automation of jobs or the application of AI technologies in cities might also be characterized by losses of certain economic activities and thus bringing in economic inequalities that bring the marginalization of opportunity to work.
Power concentration	It is a small number of large companies or governments that may possess the AI technologies, which could be a concern regarding the concentration of power and its possible use to achieve profit or political control.
Security	These AI systems can easily be subject to hacking, misuse, and threat toward public safety, privacy breaches, or manipulation of smart city systems. A very key issue is that of ownership and control of data collected by AI systems. For example, citizens might not have control over their personal data, which might be used for commercial or political purposes.
Data ownership	AI systems can further exacerbate or even create existing biases—social, racial, or economic—at the expense of already marginalized communities because data is biased or algorithmically designed with bias. Most uses of AI within Smart Cities involve huge volumes of data that breach private information if not well managed or protected. A challenge can also be in terms of the threat of surveillance.

Source: Author's own work.

the training of the system. When developers find and actively correct the biases in the data, fairer AIs will be created. To top it all, regular auditing of general AI algorithms, but most importantly sensitive applications in predictive policing and public service delivery, would be done so that deep insights could be shared related to the identification of unintended bias points accrued because of normal operations (Szpilko et al., 2023, Zamponi and Barbierato, 2022).

Data protection regulations involving consent, transparency, and minimization of data solve issues related to privacy. They shall be afforded an opportunity to decline or limit the processing of personal data, with such a mechanism clearly outlined. Besides, in Smart Cities, the AI systems are likely to be so designed that, wherever possible, data is processed locally, hence reducing the need for storage at a central facility, and thus the risk of massive data breaches. This also aligns with privacy regulations, like the General Data Protection Regulation, which set legal boundaries over personal information (Kim et al., 2021).

This shall, in return, build transparency regarding how open-source AI frameworks are applied and instill an interest by people in how AI works. It also allows the government and developers to do some needed explanation of information concerning the making of decisions with particularly high stakes in the life of citizens. These can be achieved by availing the public with discussion forums where the latter will ask questions about AI systems, give feedback, and even provide suggestions for areas needing improvements. It also creates room for building trust in the technologies by opening up flaws or biases within the systems more promptly for correction (Venumuddala et al., 2023). Accountability within AI systems could be empowered by setting clear lines of responsibility for the decision-making process. That is, developers, operators, and policymakers should show responsibility in case of an AI system's actions. These suggestions or remarks might also take into consideration possibilities of redress in case of possible damage contributed by AI. Again, ethics can be an issue in the use of AI systems within a certain set-up of oversight bodies—independent review panels that are to look into complaints. Drawing further on liability, a framework must come up to keep responsible anyone or any entity involved once failure or harm in the AI system occurs (Coppola et al., 2023; Mahrez et al., 2021).

In that respect, AI technologies for Smart Cities should be developed in a manner that caters to the needs of all citizens inclusively. It is recommended that developers work with diverse communities to understand their needs better in ensuring the AI systems developed do not create disadvantages by design for marginalized groups; this could be possible through access to AI-powered applications at cheaper costs and making sure that all classes of society stand a chance to be included in smart city innovations. More fundamentally, AI deployments should be targeted at closing gaps, not opening them—say, through specifically targeted support for disadvantaged groups (Balica and Cuțitoi, 2022).

Automation via AI is also the least destructive of lost jobs when it is pursued hand in hand with heavy investments in retraining programs and workforce development initiatives. Cities can help create collaborations among governments, businesses, and educational institutions to better prepare the workforce for success in an AI-based economy. One may have to review further alternative ways of supporting persons who will lose jobs because of automation, if not UBI or other social safety nets, so that no human is left behind in the turn of the urban environment into a place where life with the help of AI prevails (Zwitter and Helbing, 2024).

It is also applied in the concentration of power into few corporations or government entities, a factor that can be domesticated through policies seeking to ensure the operations are transparent and democratically controlled, where public entities are involved in making the decisions. Rules and regulations should be laid down to ensure that Artificial Intelligence technologies in their development and utilization are built and put into practice to serve all interests and not only some mighty actors. A public-private partnership can guarantee the creation of the AI system in a participative and inclusive manner to cater to the manifold needs of various stakeholders in a rather just way (Balashova and Reshetnikova, 2024). Concerns about AI systems in Smart Cities will be overcome by the use of appropriate cybersecurity measures that give regular updates to protect against emerging threats (Ng et al., 2023). Design the AI systems to be resilient; hence, they are able to bounce back forthwith in case of a security breach. Solutions should be developed with regular penetration tests, security audits, and monitoring to find weaknesses in the system before they are leveraged. Such collaboration among governments, technology companies, and cybersecurity experts is needed for developing common sets of cybersecurity practices that will protect both AI systems and the citizens served (Gardner, 2024).

Data ownership issues should first be approached with policy transparently and reasonably, with a focus on individual rights. This policy shall vest the ownership with the citizens, and such ownership shall be invested with a suite of complete rights over the collection, sharing, and use of this data. Data should never be misused for monetary gains on account of some commercial purposes without explicit consent, nor shall there exist such a policy that fails to ensure proper assurance of data use responsibly and ethically. Government regulatory frameworks may ensure the limitation of powers by mega-corporations in dealing with huge sums of data to make certain that handling is done transparently, securely, and with fairness (Riedmann-Streitz et al., 2024).

11.2 COLLABORATIVE APPROACHES TO AI INTEGRATION

The stakeholders want integration among the governmental agencies with the private organizations, technology firms, academia, and locals by deploying, refining, and using Artificial Intelligence systems as an approach toward assimilation of Artificial Intelligence in creation of Smart Cities for a better lifestyle (Osipova et al., 2024). This is also the kind of partnership that is very often seen coming up with new solutions for efficient, sustainable, and more livable cities where AI technologies would go hand in glove to solve multidimensional problems related to traffic congestion, resource management, environmental sustainability, and public safety (Liu et al., 2023).

Most of the collaboration models leave the lead to the governments in the determination of regulatory frameworks, data privacy policies, and strategic alignment to the bigger objectives of city development. As part of this, for example, municipalities might be working together toward the implementation of intelligent traffic systems with AI algorithms for more efficient flow, a decrease in accidents, and the least amount of emissions. These universities and research institutions bring wide-ranging expertise in data science, machine learning, and urban studies that could complement knowledge for use by AI applications with large and significant outcome orientations in a nondiscriminatory way (Gupta, 2023).

Of equal importance is the private sector involvement in novelty and technological infrastructure for large financial projects in AI. The companies that have superior algorithms in creating AI are going to apply those against city-specific solutions, be it predictive infrastructure maintenance or further optimization of waste management—possibly extending even to creating intelligent energy grids. Such cooperation with startups would mean fresh and much-needed agility in combat against those urban challenges (Moreno-Ibarra et al., 2024).

Participation by the communities at every step of the process also becomes an important factor. It is only with active participation of citizens in decision-making and collection of data that better meeting of their real needs would take place. For these reasons, such community-driven initiatives might function as a far better vehicle to distribute the advantages of integrating AI more completely, instead of widening the existing inequalities (Jeyalakshmi et al., 2024). Contributions from citizens regarding the application of AI to public services will ensure it works for and serves all types of demographic groups. The question of sharing and pooling the data is a highly relevant one in this collaborative approach. Smart Cities spew out enormous data emanating from sensors, cameras, and other IoT devices (Stecuła et al., 2023; 2024; Wolniak et al., 2024). In reality, truly impactful AI solutions need data that remains fragmented across departments and organizations. This is where active creation is in the realm of secure, interoperable data-sharing platforms, involving collaboration that would enable diverse entities to access and analyze the same data without compromising privacy and security. Therefore, full

TABLE 11.2 Key Aspects of Collaborative Approaches to AI Integration in Smart Cities

Aspect	Description
Stakeholders involved	These are governments, private sector companies, technology providers, academic institutions, local communities, startups, and NGOs.
Role of government	Establishment of regulation frameworks, data privacy, alignment of AI projects to urban development objectives, and ethical implementation of AI systems.
Role of private sector	Technology infrastructure, innovation, funding, and expertise to develop AI solutions on urban challenges such as traffic management, waste, and energy can be prepared by private sector.
Role of academia	This involves data science and machine learning research with expertise in urban studies to inform AI deployments for evidence-based solutions within cities.
Citizen involvement	It involves deep engagement with the local communities about their needs and preferences to build equity and inclusivity in how AI systems are designed and implemented locally.
Data sharing and collaboration	Data from all sectors of the industry and organizationally are shared and pooled in order to constantly upgrade the AI algorithms on resource management. Guarantee more transparency and privacy.
Innovation and pilot programs	Carry out a first pilot on AI solutions either in select areas or districts prior to scaling up to citywide applications; adopt innovation and technology to enhance them according to real data and actual feedback.
Ethical considerations	AI systems that are transparent, accountable, non-discriminatory, fair with no perpetuation of any bias or inequality; this shall be an inclusive contribution from several stakeholders in the creation of ethical guidelines
Sustainability and scalability	Utilize AI in solving environmental challenges such as energy efficiency, with scalable solutions that can expand across the city.
Public-Private Partnerships (PPP)	Joint funding and development of AI projects by governments in collaboration with private companies: sharing resources, risks, and benefits.
Interoperability and standards	Establishment of standard protocols for data exchange and interoperability of AI systems across sectors and platforms.

Source: Author's own work.

potentials are going to be realized by cooperation in availing full datasets, which are needed for furthering machine learning models toward making better predictions and suitable allocation of resources (Zhang et al., 2022; Hsu, 2018).

The former situation is very important in order to achieve innovation and scalability for the AI integration approach within collaborative smart city operation processes (Wolniak et al., 2024; Wolniak and Stecuła, 2024a). Third-party stakeholders can do pilots on the programs running at a few areas before scaling them up. A concrete example is that all the districts of the town first deploy and refine only the energy efficiency solutions employing the AIs, then deploy others later, for scaling up (Gajdzik et al., 2023; 2024; Nagaj et al., 2024). Iterative by nature, the concept allows for continuous improvement to address real-world problems and needs, evolving responsively. (Pahuja, 2022).

The other important insight into ethical grounds on which the collaborative approaches stand most strongly is that with rising embeddedness within the urban landscape, there comes also a looming imperator of collaboration on increasing demands for transparency, accountability, and fairness. These again are some of the key motives for which stakeholders strive so that AI applications do not institutionalize biases or fuel the growing inequalities. These, coupled with the collaborations, have been the pathways to laying down ethical guidelines and governance frameworks pertaining to such concerns and infuse public confidence to make sure of responsible use of AI (Lane, 2023; Shulajkovska et al., 2024). The summarized version of the main important aspects of the collaborative approaches toward integrating AI into a smart city is in Table 11.2.

11.3 VISION FOR THE FUTURE: INCLUSIVE AND SUSTAINABLE SMART CITIES

The use of AI technologies integrated within urban planning and management will meet a wide range of different challenges that diverse populations have, including people with disabilities and marginalized groups, in such a way as to narrow the gap within disparate socio-economic segments (Acevedo and Duque, 2021; Wolniak and Stecuła, 2024b). In addition, AI will also contribute to inclusivity within Smart Cities by developing intelligent transportation systems. On one way, AI-driven systems enable enhancement of the networks so that public transport can more easily be made available in larger parts to persons with disabilities, elderly citizens, or inhabitants of less privileged areas in metropolitan cities (Wolniak and Skotnicka-Zasadzień, 2018, 2021). For example, it may design the route so that less accessible areas may be connected with a primary center for moving into a city. Moreover, AI detects traffic patterns in real time and changes to service, which in turn limit congestion and create fairer access to transportation (Bleja et al., 2020; Bosch et al., 2021; Kim et al., 2018; László and Ráhel, 2016).

Artificial Intelligence also contributes a great deal to building better health mechanisms in cities. Using Artificial Intelligence-enabled sensors and wearable technology, data analytics provides evidence for cities to bring customized health-related solutions to those groups who are underprivileged and also support real-time identification of health risks, thereby providing opportunities for disease prediction so as to enable timely interventions, which may prove significant in improving public health (Park, 2024). Besides, AI can be applied to optimize resources in healthcare, being equitably distributed in order to ensure that medical services will be offered and healthcare facilities are better equipped to meet citizens' needs, independent of geographical distances or social status (Bolay, 2020; Oviedo, 2021; Svelec et al., 2020).

AI in urban development and housing should be inclusive, affordable, and sustainable. With consideration of the needs in housing, impacts on the environment, and resources, AI is going to be supportive in bringing forth solutions to each need through the planners involved. In the building of new structures and infrastructures that cater to persons with special needs, it will be possible to use AI even in the most sensitive constructions (Alizadeh and Sharifi, 2023; Kolotouchkina et al., 2022). That can similarly apply in observing conditions around the living space of environments in which they exist and performing changes in real time to attain better settings; for example, air and light adjusting (Torres-González et al., 2023; Kim, 2023; Kundu, 2020).

It can be used educationally in creating personalized learning environments whereby every citizen is guaranteed an appropriate educational opportunity, irrespective of his or her age or background. It can also offer adaptive learning technologies through AI-powered platforms fitted to the needs of every particular student, even those with learning disabilities or other problems. It means inclusiveness within schools, but also in life-long learning—that is, permitting individuals to continue having an education that will keep them abreast of a rapidly changing employment market and technological change (Borghetti et al., 2021; Jayashree et al., 2019; Nápoles et al., 2021).

Artificial Intelligence stands tall for immense potential in implementation to improve community participatory governance at the local levels. Thus, the capabilities developed through AI-powered civic engagement can facilitate a more representative participation of citizens, amplifying the voices of under-represented minorities . In a nutshell, the potential with AI is that it helps to ensure public services are delivered effectively on account of proper resource allocations done in response to the real needs of the communities (Alidadi and Sharifi, 2022; Groth, 2019; Mello Rose et al., 2022; Vinod Kumar et al., 2021; Herrschel., 2017).

AI can contribute a lot to the sustainability factors of Smart Cities, mainly related to improving the quality of life of citizens (Jonek-Kowalska and Wolniak, 2023; Jonek-Kowalska, 2022a). Its focus is on energy consumption optimization and waste reduction by implementing sustainable management (Wolniak and Skotnicka-Zasadzień, 2023; Jonek-Kowalska, 2022b; Trzeciak and Jonek-Kowalska, 2021). These would be further supported by continuous analyses in that it would enable the cities to reach a balance between demand over needs from the point of view of lighting, heating, or cooling, with minimum wastage and without compromising comfort and safety (Lamonaca and Batel, 2023; McKenna, 2017). This negative environmental impact, if AI reduces it, then avails an opportunity for all citizens to have a healthy urban environment, free of pollution—so basic and foundational to inclusivity in this rapidly changing world of ours. Table 11.3 depicts different ways by which AI may support making the smart city inclusive, sustainable, and efficient in all respects for every section of society (Mboup et al., 2017; Valdez et al., 2020; Bordini et al., 2020; Bunyan and Collins, 2013).

TABLE 11.3 Benefits of AI Implementation in Smart Cities

Benefit	Description
Improved public transportation	It will provide much-needed route optimization, decrease congestion, and make systems available to all, including the elderly and disabled, furthering equity in transportation.
Enhanced healthcare services	AI can offer personalized healthcare to forecast health risks and optimize resources for the best health outcome of all citizens.
Sustainable resource management	AI helps in optimization of energy consumption, water use, and waste handling in cities for better ecological sustainability with reduced generation of waste.
Smart housing solutions	AI can support design for available, accessible, and sustainable housing, making urban development inclusive.
Personalized education	AI can offer personalized learning to all citizens, considering their various learning needs and making lifelong learning a reality.
Efficient public services	AI can make the processes of city services with regard to waste collection, law enforcement, and emergency response more effective and timely in terms of service delivery and equity.
Improved safety and security	AI-powered surveillance and predictive policing help in identifying risks in real time for the safety of all of its inhabitants.
Inclusive civic engagement	AI-driven platforms make for wider, more inclusive civic participation wherein their voice is represented in policymaking.
Environmental monitoring and control	AI can continuously monitor air quality, noise levels, and other environmental factors to intervene on time and keep a healthy urban environment.
Economic empowerment	AI can enable economic opportunities available to all citizens through better job matching, support for small businesses, and skills development programs.
Traffic management and optimization	AI can predict patterns in traffic, adjust signals to keep traffic flowing, and reduce congestion on the roads by optimizing efficiency and improving mobility.
Smart grid management	AI can optimize energy grids for better supply and demand balancing, integrate renewable energy sources, and reduce energy waste.
Disaster and emergency response	AI can predict the occurrence of a natural disaster, can co-ordinate responses in an emergency, and optimize responses to reduce the amount of damages and save lives during disasters.
Public health surveillance	AI enables tracking infectious outbreaks and analyzes the trends so as to aid cities in their faster reaction against health threats for reducing infectious diseases.
Improved waste management	AI will optimize the schedules of waste collection, reduce landfill usage, and improve recycling processes for cleaner, greener urban areas.
Smart water management	AI detects leaks and can predict water demand in order to manage resources with the least waste of available resources.
Better urban planning	AI analyzes the growth trend in population, usage of resources, and development of land to support planners with better and more sustainable planning of the urban environment.
Integration of renewable energy	AI integrates renewable sources of energy—solar and wind energies—along the grid to ensure energy sustainability by moving away from reliance on fossil fuels.
Predictive maintenance for infrastructure	AI can predict when roads, bridges, and utilities need to be maintained. The advantage is cost reduction and avoidance of major failures.
Remote monitoring and management	AI-powered IoT devices can also monitor the infrastructure of a city remotely in real time for quicker responses and efficiency in maintenance.

Source: Author's own work.

BIBLIOGRAPHY

Acevedo, E.A.C., and Duque, A.J. (2021). Guidelines to define a regulatory proposal in the transition and inclusion of non-conventional renewable energies in Colombia and its role in the development of smart cities. *Smart Innovation, Systems and Technologies*, 233, 227–237.

Alidadi, M., and Sharifi, A. (2022). The extent of inclusion of smart city indicators in existing urban sustainability assessment tools. In: B. Russo, & F.A.L. Pacheco (Eds.), *Urban Climate Adaptation and Mitigation*. Amsterdam, Netherlands: Elsevier, pp. 175–198.

Alizadeh, H., and Sharifi, A. (2023). Toward a societal smart city: Clarifying the social justice dimension of smart cities. *Sustainable Cities and Society*, 95, 104612.

Arora, A., Jain, A., Yadav, D., Hassija, V., Chamola, V., and Sikdar, B. (2023). Next generation of multi-agent driven smart city applications and research paradigms. *IEEE Open Journal of the Communications Society*, 4, 1–10. https://doi.org/10.1109/OJCOMS.2023.3245678

Balashova, S.A., and Reshetnikova, M.S. (2024). Business sector and the smart city in the artificial intelligence (AI) era: Ethics, privacy, and cybersecurity. In: A. Visvizi, O. Troisi, & M. Grimaldi (Eds.), *ICT, the Business Sector and Smart Cities*. Abingdon, Oxfordshire, United Kingdom: Routledge, pp. 210–226.

Balica, R.-Ş., and Cuţitoi, A.-C. (2022). Ethical artificial intelligence in smart mobility technologies: Autonomous driving algorithms, geospatial data mining tools, and ambient sound recognition software. *Contemporary Readings in Law and Social Justice*, 14(1), 64–81. https://doi.org/10.22381/crlsj1420225

Bleja, J., Langer, H., Grossmann, U., and Morz, E. (2020). Smart cities for everyone-age and gender as potential exclusion factors. *2020 IEEE European Technology and Engineering Management Summit, E-TEMS* 2020, 9111741.

Bolay, J.-C. (2020). When inclusion means smart city: Urban planning against poverty. *Advances in Intelligent Systems and Computing*, 1069, 283–299.

Bordini, R.H., Mascardi, V., Costantini, S., Lespérance, Y., Ricci, A. (2020). Transcultural health-aware guides for the elderly. *CEUR Workshop Proceedings*, 2706, 135–146.

Borghetti, F., Longo, M., Mazzoncini, R., Cesarini, L., and Contestabile, L. (2021). Relationship between railway stations and the territory: Case study in Lombardy – Italy for 15-min station. *International Journal of Transport Development and Integration*, 5(4), 367–378.

Bosch, E.R., Wybraniec, B., Lazzarini, B., Junyent, M.V., and Crespo, À.G. (2021). The DIGNITY Project: Toward a system of inclusive digital mobility in the Barcelona metropolitan area. *Transportation Research Procedia*, 58, 134–141.

Bunyan, S., and Collins, A. (2013). Digital exclusion despite digital accessibility: Empirical evidence from an English City. *Tijdschrift voor Economische en Sociale Geografie*, 104(5), 588–603.

Coppola, G., Varde, A.S., and Shang, J. (2023). Enhancing cloud security posture for ubiquitous data access with a cybersecurity framework based management tool. In *Proceedings of the 2023 IEEE 14th Annual Ubiquitous Computing, Electronics & Mobile Communication Conference (UEMCON)* (pp. 590–594). IEEE. https://doi.org/10.1109/UEMCON2023.10316003

Darmawan, A.K., Muhsi, M., Anekawati, A., Sakdiyah, H., Yusuf, M., Sophan, M.K., Ferdiansyah, D., Umam, B.A., and Jalil, D.K.A. (2023). An interpretive structural model approach to strategic management modelling for sustainable smart village development in Indonesia. In *Proceedings of the 2023 10th International Conference on ICT for Smart Society (ICISS)* (pp. 1–7). IEEE. https://doi.org/10.1109/ICISS59129.2023.10291310

Dong, L., and Liu, Y. (2023). Frontiers of policy and governance research in a smart city and artificial intelligence: An advanced review based on natural language processing. *Frontiers in Sustainable Cities*, 5, 1199041. https://doi.org/10.3389/frsc.2023.1199041

Duan, Y., Xu, C., Hu, W., and Zhao, W. (2022). Research on the regulation of artificial intelligence on the orderly development of China's intelligent economy. In *Proceedings of the 2022 Asia-Pacific Computer Technologies Conference (APCT)* (pp. 11–15). IEEE. https://doi.org/10.1109/APCT55107.2022.00017

Gajdzik, B., Awdziej, M., Jaciow, M., Lipowska, I., Lipowski, M., Szojda, G., Tkaczyk, J., Wolniak, R., Wolny, R., and Grebski, W.W. (2024). Encouraging residents to save energy by using smart transportation: Incorporating the propensity to save energy into the UTAUT model. *Energies*, 17(21), 5341. https://doi.org/10.3390/en17215341

Gajdzik, B., Bartuś, K., Jaciow, M., Wolniak, R., Wolny, R., and Grebski, W.W. (2024). Evolution of Polish e-consumers' environmental awareness and purchasing behavior over ten years. *Sustainability*, 16(11), 4686. https://doi.org/10.3390/su16114686

Gajdzik, B., Wolniak, R., Nagaj, R., Grebski, W.W., and Romanyshyn, T. (2023). Barriers to Renewable Energy Source (RES) installations as determinants of energy consumption in EU countries. *Energies*, 16(21), 7364. https://doi.org/10.3390/en16217364

Gajdzik, B., Wolniak, R., Nagaj, R., Žuromskaitė-Nagaj, B., and Grebski, W.W. (2024). The influence of the global energy crisis on energy efficiency: A comprehensive analysis. *Energies*, 17(4), 947. https://doi.org/10.3390/en17040947

Gardner, N. (2024). Scaling the smart city: The design and ethics of urban technology, scaling the smart city. In: *The Design and Ethics of Urban Technology*. Amsterdam, Netherlands: Elsevier, pp. 1–194.

Groth, S. (2019). Multimodal divide: Reproduction of transport poverty in smart mobility trends. *Transportation Research Part A: Policy and Practice*, 125, 56–71.

Gupta, S. (2023). Mapping the Smart City: Participatory approaches to XAI. *DIS 2023 Companion: Companion Publication of the 2023 ACM Designing Interactive Systems Conference*, pp. 1–6.

Helbing, D., Fanitabasi, F., Giannotti, F., Pedreschi, D., and Pournaras, E. (2021). Ethics of smart cities: Towards value-sensitive design and co-evolving city life. *Sustainability*, 13(20), 11162.

Herrschel, T. (2017). Metropolitan elitism, marginalisation and the need for 'smartness' in governing the regionalised city. *Territorio*, (83), 32–36.

Hsu, K.K. (2018). AI city's solar energy collaborative commerce and sharing economy. *Proceedings of the International Conference on Electronic Business (ICEB), 2018-December*, pp. 98–105.

Jayashree, P., Hamza, F., El Barachi, M., and Gholami, G. (2019). Inclusion as an enabler to sustainable innovations in smart cities: A multi-level framework. *2019 4th International Conference on Smart and Sustainable Technologies*, SpliTech 2019, 8783013.

Jeyalakshmi, J., Vijay, K., Berna, E., and Manohar, E. (2024). Smart cities and urban sustainability with AI. In: R.C. Rajappan, N.S.G. Ganesh, & J.A. Daniel (Eds.), *The Convergence of Self-Sustaining Systems with AI and IoT*. Pennsylvania, USA: IGI Global, pp. 330–350.

Jonek-Kowalska, I. (2022a). Assessing the energy security of European countries in the resource and economic context. *Oeconomia Copernicana*, 13(2), 301–334. https://doi.org/10.24136/oc.2022.011

Jonek-Kowalska, I. (2022b). Housing infrastructure as a determinant of quality of life in selected Polish smart cities. *Smart Cities*, 5(3), 924–946. https://doi.org/10.3390/smartcities5030046

Jonek-Kowalska, I., and Wolniak, R. (2023). *Towards Sustainability and a Better Quality of Life?* Routledge: London, UK.

Jonek-Kowalska, I., and Wolniak, R. (2025). *Exclusion in Smart Cities Assessment and Strategy for Strengthening Resilience*. Routledge: London, UK.

Kim, D., Kwon, H.-Y., Jun, D., Hagen, L., and Chun, S.A. (2018). Opportunities and challenges in the intelligent society: Smart cities, digital inclusion, and cybersecurity. *ACM International Conference Proceeding Series*, a131.

Kim, K. (2023). Inclusion, exclusion, and participation in digital polis: Double-edged development of poor urban communities in alternative smart city-making. In: B. Dahiya (Ed.), *Advances in 21st Century Human Settlements*. Cham, Switzerland: Springer Nature, pp. 155–178.

Kim, K., Kim, J.S., Jeong, S., Park, J.-H., and Kim, H.K. (2021). Cybersecurity for autonomous vehicles: Review of attacks and defense. *Computers & Security*, 103, 102150. https://doi.org/10.1016/j.cose.2020.102150

Kolotouchkina, O., Barroso, C.L., and Sánchez, J.L.M. (2022). Smart cities, the digital divide, and people with disabilities. *Cities*, 123, 103613.

Kundu, D. (2020). Political and social inclusion and local democracy in Indian cities: Case studies of Delhi and Bengaluru. In: *Advances in 21st Century Human Settlements*. Singapore: Springer Nature, pp. 185–208.

Lamonaca, L., and Batel, S. (2023). What of recognition justice? An empirical analysis of the role of recognition justice in social housing smart city projects in the Global North. In: P. Velasco-Herrejón, B. Lennon, & N.P. Dunphy (Eds.), *Living with Energy Poverty Perspectives from the Global North and South*. London, United Kingdom: Routledge, pp. 244–254.

Lane, L. (2023). Preventing long-term risks to human rights in smart cities: A critical review of responsibilities for private AI developers. *Internet Policy Review*, 12(1), 1–28.

László, G., and Ráhel, C. (2016). Do smart cities intensify social exclusion?, *Informacios Tarsadalom*, 16(3), 83–100.

Liu, H.K., Guo, Y.-W., and Chen, L.-Y. (2023). Crowdsourcing Smart City: SmartTaipei Project. *ACM International Conference Proceeding Series*, pp. 44–54.

Mahrez, Z., Sabir, E., Badidi, E., Saad, W., and Sadik, M. (2021). Smart urban mobility: When mobility systems meet smart data. *IEEE Transactions on Intelligent Transportation Systems*, 23(6), 6222–6239. https://doi.org/10.1109/TITS.2021.3084907

Mboup, G., Diongue, M., and Ndiaye, S. (2017). Smart city foundation—driver of smart cities. In: A. Kirby, E. Friedberg, R.P.B. Singh, K. Yu, M.El Sioufi, T. Campbell, Y. Hayashi, X. Bai, & D. Haase (Eds.), *Advances in 21st Century Human Settlements*. Singagore: Springer, pp. 841–869.

McKenna, H.P. (2017). Re-conceptualizing social inclusion in the context of 21st-century smart cities. In: J. Choudrie, S. Kurnia, & P. Tsatsou (Eds.), *Social Inclusion and Usability of ICT-Enabled Services*. Cham: Springer, pp. 49–66.

Mello Rose, F., Thiel, J., and Grabher, G. (2022). Selective inclusion: Civil society involvement in the smart city ecology of Amsterdam. *European Urban and Regional Studies*, 29(3), 369–382.

Momen, N. (2023). Privacy and ethics in a smart city: Towards attaining digital sovereignty. In: A.J.M. Gisela Bichler, T. Bourlai, C. Johnson, P. Karampelas, C. Leuprecht, E.C. Morse, & Y. Yamagata (Eds.), *Advanced Sciences and Technologies for Security Applications*. Cham: Springer, pp. 47–60.

Moreno-Ibarra, M., Saldaña-Perez, M., Rodríguez, S.P., and Carbajal, E.J. (2024). Generative AI (GenAI) and smart cities: Efficiency, cohesion, and sustainability. In: A.V., &

H. Godlewska-Majkowska (Eds.), *Smart Cities: Lock-in, Path-dependence and Non-linearity of Digitalization and Smartification*. Abingdon, Oxon, Great Brittan: Routledge, pp. 118–129.

Nagaj, R., Gajdzik, B., Wolniak, R., and Grebski, W.W. (2024). The impact of deep decarbonization policy on the level of greenhouse gas emissions in the European Union. *Energies*, 17(5), 1245. https://doi.org/10.3390/en17051245

Nápoles, V.M.P., Páez, D.G., Penelas, J.L.E., de Pablos, F.M., and Gil, R.M. (2021). *Social inclusion in smart cities*. In: J.C. Augusto (Ed.), *Handbook of Smart Cities*. Cham: Springer, pp. 469–514.

Ng, L.H.X., Lim, A.C.M., Lim, A.X.W., Taeihagh, A. (2023). Digital ethics for biometric applications in a smart city. *Digital Government: Research and Practice*, 4(4), 26.

Osipova, M., Tost, J., Buruk, O.O., Bardzell, J., and Hornecker, E. (2024). AI as a citizen: Eliciting smart city future stories through human-AI collaborative fiction writing. *ACM International Conference Proceeding Series*, pp. 330–335.

Oviedo, D. (2021). Making the links between accessibility, social and spatial inequality, and social exclusion: A framework for cities in Latin America. *Advances in Transport Policy and Planning*, 8, 135–172.

Pahuja, N. (2022). Partnering with technology firms to train smart city workforces. In: J.R. Vacca (Ed.), *Smart Cities Policies and Financing: Approaches and Solutions*. Amsterdam: Elsevier, pp. 169–180.

Park, I.K. (2024). Unravelling the relationship between spatial and social inclusion: Evidence from Korean cities and regions. *Local Environment*, 29(1), 21–39.

Phillips, C., and Jiao, J. (2023). Artificial intelligence & smart city ethics: A systematic review. *2023 IEEE International Symposium on Ethics in Engineering, Science, and Technology: Ethics in the Global Innovation Helix, ETHICS 2023*.

Riedmann-Streitz, C., Streitz, N., Antona, M., Rau, P.-L.P., and Rosenzweig, E. (2024). How to create and foster sustainable smart cities? Insights on ethics, trust, privacy, transparency, incentives, and success. *International Journal of Human-Computer Interaction*, 41(4), 2491–2522.

Sawhney, N. (2023). Contestations in urban mobility: Rights, risks, and responsibilities for urban AI. *AI & Society*, 38(4), 1083–1098. https://doi.org/10.1007/s00146-022-01502-2

Schintler, L.A., and McNeely, C.L. (2022). Artificial intelligence, institutions, and resilience: Prospects and provocations for cities. *Journal of Urban Management*, 11(3), 256–268. https://doi.org/10.1016/j.jum.2022.07.004

Shulajkovska, M., Smerkol, M., Noveski, G., and Gams, M. (2024). Enhancing urban sustainability: Developing an open-source AI framework for smart cities. *Smart Cities*, 7(5), 2670–2701.

Skubis, I., Wolniak, R., and Grebski, W.W. (2024). AI and human-centric approach in smart cities management: Case studies from Silesian and lesser Poland voivodships. *Sustainability*, 16(18), 8279.

Stecuła, K., Wolniak, R., and Aydın, B. (2024). Technology development in online grocery shopping—from shopping services to virtual reality, metaverse, and smart devices: A review. *Foods*, 13(23), 3959.

Stecuła, K., Wolniak, R., and Grebski, W.W. (2023). AI-driven urban energy solutions—from individuals to society: A review. *Energies*, 16, 7988.

Svelec, D., Bjelcic, N., and Blazekovic, M. (2020). Smart cities as an opportunity and challenge for people with disabilities. *2020 43rd International Convention on Information, Communication and Electronic Technology*, MIPRO 2020 - Proceedings, 456–461, 9245183.

Szpilko, D., Naharro, F.J., Lăzăroiu, G., Nica, E., and de la Torre Gallegos, A. (2023). Artificial intelligence in the smart city—A literature review. *Engineering Management in Production and Services*, 15(1), 53–75. https://doi.org/10.2478/emj-2023-0005

Tao, M., Jiang, R., and Downs, C. (2022). Ethics of face recognition in smart cities toward trustworthy AI. In: A.J. Masys, G. Bichler, T. Bourlai, C. Johnson, P. Karampelas, C. Leuprecht, E.C. Morse, & Y. Yamagata (Eds.), *Advanced Sciences and Technologies for Security Applications*. Cham: Springer, pp. 23–52.

Torres-González, M.A., Hernández-Veleros, Z.S., Ramírez-Rosas, J.G., and Guzmán-Escorza, L.E. (2023). Older adults in Mexico: Their inclusion in ICTs and smart cities. In: J.R. Vanoye (Ed.), *Management, Technology, and Economic Growth in Smart and Sustainable Cities*. Cham: Springer, pp. 111–132.

Trzeciak, M., and Jonek-Kowalska, I. (2021). Monitoring and control in program management as effectiveness drivers in Polish energy sector: Diagnosis and directions of improvement. *Energies*, 14(15), 4661. https://doi.org/10.3390/en14154661

Valdez, A.-M., Wigley, E., Zanetti, O., and Gillian, G. (2020). Learning lessons for avoiding the inadvertent exclusion of communities from smart city projects. In: A. Aurigi, & N. Odendaal (Eds.), *Shaping Smart for Better Cities: Rethinking and Shaping Relationships between Urban Space and Digital Technologies*. Cham: Springer, pp. 221–237.

van Gils, B.A.M., and Bailey, A. (2023). Revisiting inclusion in smart cities: Infrastructural hybridization and the institutionalization of citizen participation in Bengaluru's peripheries. *International Journal of Urban Sciences*, 27(S1), 29–49.

Venumuddala, V.R., Prakash, A., and Chaudhuri, B. (2023). Assessment of smart city IoT interventions: A complexity informed ethics perspective. *International Conference on Information Systems, ICIS 2023: "Rising like a Phoenix: Emerging from the Pandemic and Reshaping Human Endeavors with Digital Technologies"*.

Vinod Kumar, T.M., Sruthi Krishnan, V., Deepak Lawrence, K., Mohammed Firoz, C., and Cyriac, S. (2021). Design of smart global economic community in Kattangal. In: B. Dahiya (Ed.), *Advances in 21st Century Human Settlements*. Singapore: Springer, pp. 337–420.

Wolniak, R., Gajdzik, B., Grebski, M., Danel, R., and Grebski, W.W. (2024). Business models used in smart cities—theoretical approach with examples of smart cities. *Smart Cities*, 7, 1626–1669.

Wolniak, R., and Jonek-Kowalska, I. (2022). The creative services sector in Polish cities. *Journal of Open Innovation: Technology, Market, and Complexity*, 8, 17.

Wolniak, R., and Skotnicka-Zasadzień, B. (2018). Developing a model of factors influencing the quality of service for disabled customers in the condition s of sustainable development, illustrated by an example of the Silesian voivodeship public administration. *Sustainability*, 10(7), 2171, 1–17.

Wolniak, R., and Skotnicka-Zasadzień, B. (2021). Improvement of services for people with disabilities by public administration in Silesian Province Poland. *Sustainability*, 13(2), 967, 1–16.

Wolniak R., Skotnicka-Zasadzień, B., and Zasadzień M. (2019). Problems of the functioning of e-administration in the Silesian region of Poland from the perspective of a person with disabilities. *Transylvanian Review of Administrative Sciences*, 57E, 137–155,

Wolniak, R., and Skotnicka-Zasadzień, B. (2023). Development of wind energy in EU countries as an alternative resource to fossil fuels in the years 2016–2022. *Resources*, 12(8), 96. https://doi.org/10.3390/resources12080096

Wolniak, R., and Stecuła, K. (2024a). Artificial intelligence in smart cities—Applications, barriers and future directions: A review. *Smart Cities*, 7, 1346–1389.

Wolniak, R., and Stecuła, K. (2024b). Evaluation of quality of innovative e-learning in higher education: An insight from Poland. *Applied System Innovation*, 7(6), 109. https://doi.org/10.3390/asi7060109

Wolniak, R., Stecuła, K., and Aydın, B. (2024). Digital transformation of grocery in-store shopping-scanners, artificial intelligence, augmented reality and beyond: A review. *Foods*, 13(18), 2948.

Zamponi, M.E., and Barbierato, E. (2022). The dual role of artificial intelligence in developing smart cities. *Smart Cities*, 5(3), 728–755. https://doi.org/10.3390/8

Zhang, Y., Zong, R., Shang, L., Zeng, H., and Wang, D. (2022). CrowdOptim: A crowd-driven neural network hyperparameter optimization approach to AI-based smart urban sensing. *Proceedings of the ACM on Human-Computer Interaction*, 6(CSCW2), 435.

Zwitter, A., and Helbing, D. (2024). Ethics of smart cities and smart societies. *Ethics and Information Technology*, 26(4), 69.

Index